Lecture Notes in Physics

New Series m: Monographs

Managing Editor

W. Beiglböck
Assisted by Mrs. Sabine Landgraf
c/o Springer-Verlag, Physics Editorial Department II
Tiergartenstrasse 17, D-69121 Heidelberg, Germany

The Editorial Policy for Monographs

The series Lecture Notes in Physics reports new developments in physical research and teaching - quickly, informally, and at a high level. The type of material considered for publication in the New Series m includes monographs presenting original research or new angles in a classical field. The timeliness of a manuscript is more important than its form, which may be preliminary or tentative. Manuscripts should be reasonably self-contained. They will often present not only results of the author(s) but also related work by other people and will provide sufficient motivation, examples, and applications.

The manuscripts or a detailed description thereof should be submitted either to one of the series editors or to the managing editor. The proposal is then carefully refereed. A final decision concerning publication can often only be made on the basis of the complete manuscript, but otherwise the editors will try to make a preliminary decision as definite as they can on the basis of the available information.

Manuscripts should be no less than 100 and preferably no more than 400 pages in length. Final manuscripts should preferably be in English, or possibly in French or German. They should include a table of contents and an informative introduction accessible also to readers not particularly familiar with the topic treated. Authors are free to use the material in other publications. However, if extensive use is made elsewhere, the publisher should be informed. Authors receive jointly 50 complimentary copies of their book. They are entitled to purchase further copies of their book at a reduced rate. As a rule no reprints of individual contributions can be supplied. No royalty is paid on Lecture Notes in Physics volumes. Commitment to publish is made by letter of interest rather than by signing a formal contract. Springer-Verlag secures the copyright for each volume.

The Production Process

The books are hardbound, and quality paper appropriate to the needs of the author(s) is used. Publication time is about ten weeks. More than twenty years of experience guarantee authors the best possible service. To reach the goal of rapid publication at a low price the technique of photographic reproduction from a camera-ready manuscript was chosen. This process shifts the main responsibility for the technical quality considerably from the publisher to the author. We therefore urge all authors to observe very carefully our guidelines for the preparation of camera-ready manuscripts, which we will supply on request. This applies especially to the quality of figures and halftones submitted for publication. Figures should be submitted as originals or glossy prints, as very often Xerox copies are not suitable for reproduction. For the same reason, any writing within figures should not be smaller than 2.5 mm. It might be useful to look at some of the volumes already published or, especially if some atypical text is planned, to write to the Physics Editorial Department of Springer-Verlag direct. This avoids mistakes and time-consuming correspondence during the production period.

As a special service, we offer free of charge LaTeX and TeX macro packages to format the text according to Springer-Verlag's quality requirements. We strongly recommend authors to make use of this offer, as the result will be a book of considerably improved technical quality.

Manuscripts not meeting the technical standard of the series will have to be returned for improvement.

For further information please contact Springer-Verlag, Physics Editorial Department II, Tiergartenstrasse 17, D-69121 Heidelberg, Germany.

Allan J. Greer   William J. Kossler

Low Magnetic Fields in Anisotropic Superconductors

Springer

Authors

Allan J. Greer
Physics Department
Washington and Jefferson College
60 South Lincoln Street
Washington, PA 15301-4801, USA

William J. Kossler
Physics Department
College of William and Mary
Williamsburg, VA 23187-8795, USA

ISBN 3-540-59167-2 Springer-Verlag Berlin Heidelberg New York

CIP data applied for.

© Springer-Verlag Berlin Heidelberg 1995
Printed in Germany

Typesetting: Camera-ready by the author
SPIN: 10127359   55/3142-543210 - Printed on acid-free paper

Contents

Chapter 1

Introduction

Superconductors have been known about since the turn of the century. Recently there has been a renewed interest with the discovery of the new, high-T_c materials since 1986[1]. These compounds become superconducting at much warmer temperatures than any previously known. In fact, many of tthem superconduct at temperatures above the boiling point of liquid nitrogen, making the observation of the transition both accessible and inexpensive. It was obvious immediately that these materials could have a tremendous technological impact, or lead to further materials with even higher transitions. For this reason there has been an intense effort by scientists in both academia and industry to study these materials. The scientific and industrial communities hope to learn what makes these materials work. For, learning how these materials work not only increases mankind's overall knowledge of his world, but could make some person or company quite successful if the information were used and developed correctly.

This work is a small part of the scientific community's efforts to better understand the high-T_c materials. In particular, it is a theoretical and numerical study of anisotropic superconductors using the techniques of muon spin rotation spectroscopy. A phenomenological theory called the London theory is used in its anisotropic form to describe the magnetic fields which can exist within certain kinds of these materials. Once these fields are known, a numerical simulation can be performed by allowing muons to stop uniformly within these fields. The stopped muons' behavior is governed by the local magnetic environment within the superconductor, and their subsequent decay into positrons can be modeled numerically, also. It is therefore possible to predict what detectors would see if muons were to stop within such a magnetic field distribution. The resulting simulated data is useful in itself, but with further manipulation can be used to obtain a variety of information about the microscopic magnetic fields which have influenced the muons' behavior.

The discussion proceeds in the following way. First, a brief introduction to the fundamentals of muon spin rotation (μSR) is given. This includes where and how muons are produced, how they are transported from the production target to the sample, and how they stop in this sample. The third chapter describes how muons behave in the presence of a magnetic field, as well as the μSR techniques – time differential, transverse field, longitudinal field, and zero field.

In the fourth chapter we switch gears and delve into the world of superconductivity. Here a brief introduction to the subject is given: penetration depth, coherence length, and type of superconductor are discussed. The high temperature superconductor $YBa_2Cu_3O_{7-\delta}$ (YBCO) is described in some detail because there is much information about its magnetic behavior available and because it should be properly described by the London theory. The phenomenological London theory, which is often used for the calculation of the microscopic magnetic fields in this type of material, is developed in both the isotropic and anisotropic forms. The anisotropic case is more fully developed because many of the high-Tc materials (including YBCO) tend to be anisotropic. In this case, there arises fields which are transverse to the average magnetic field direction. These fields are the subject of much study later, in chapter 5. The anisotropic theory is also presented as a prescription for the numerical calculation of the fields at any point within the superconductor.

Chapter five describes the implementation of this prescription and its results. The results, which are calculated for parameters close to those of YBCO, are in the form of magnetic field surfaces, contours, and distributions. A study of the fields (including the transverse components mentioned above) as functions of various parameters is presented both for interest and as a check on the proper working of the program.

Chapter six develops a procedure for using the fields calculated in chapter five to do a numerical simulation of muons stopping in the superconductor. The results of the simulation give information about the fields experienced by the muons after they stop. Further, the simulation results are then used in a novel extension of the usual μSR techniques to yield even more information about the fields. We can obtain what we call *moments* of the magnetic field distribution, which can tell us about the transverse fields mentioned above. In addition, we can get a handle on the direction of the average internal magnetic field $\mathbf{B}$, which in general is not in the same direction as the applied field $\mathbf{H}_a$.

Chapter 2

Fundamentals of μSR

This chapter briefly discusses some of the basic concepts and background of μSR. Topics such as the production of pions, the production of muons from these pions, and the stopping of the muons in a sample are described. The two major types of beam lines are introduced, and the kinematics of the particles in each type are derived. Finally, the effects of stopping on the muon polarization are discussed.

2.1   A Little History

In 1957 the groups consisting of Garwin, Lederman, and Weinrich[2] and Friedman and Telegdi[3], working independently and concurrently verified the theory of Lee and Yang [4, 5] of the failure of conservation of parity and charge conjugation in weak decays. These experiments were carried out by studying the decays of positive and negative muons. It was found, among other things, that the positive muon had a rather large decay asymmetry – *i.e.* it decays preferentially in the direction of its spin. This led to the suggestion that muons could be used to probe internal magnetic fields in materials. Thus μSR was born.

2.2 What is μSR?

μSR stands for Muon Spin Rotation, Relaxation, or Resonance. The different R's correspond to different uses of muons. Rotation means the precession of the muon's spin about the local magnetic field where it sits. Relaxation refers to the spin-lattice relaxation rate (T_1) and the spin-spin relaxation (depolarization) rate (T_2) which can be measured with various μSR techniques. Resonance is analogous to NMR, where a radio frequency (RF) electromagnetic field is applied to the sample, causing transitions between the hyperfine energy levels of the muon. For the proper field values this reduces the decay asymmetry and thus allows the determination of the hyperfine energy levels at the muon's local environment.

The basic μSR procedure is as follows, described with reference to Fig. 2.1, which was the proposed μSR beam line for Brookhaven National Laboratory in 1990-1991. Positive muons are produced at or near a target T (Platinum, in this case) in a beam line in one of a variety of ways discussed below. The beam line usually has a bend or two (B1 and B2), as well as a separator or two (not shown), to allow the selection of muons from other particles which may be in the beam line. After selection, the muons are magnetically focused into a small spot at the target, labeled TGT in the figure. (There is much equipment associated with the target apparatus not shown here. It will be discussed later below.) Once stopped in the target, a muon's behavior will be governed by the local magnetic (and possibly thermal) environment within the material. At some later point ($0 - 10 \ \mu s$) the muon decays into a positron (e^+) and two neutrinos (ν). The positron is detected and the muon's life-time is scored in a histogram. Analysis of the time histograms yields information about the target material, as will be discussed in some detail below.

This work will deal almost exclusively with muon spin rotation, but readers interested in the other techniques, as well as more on rotation, can find information in references [6, 7, 8].

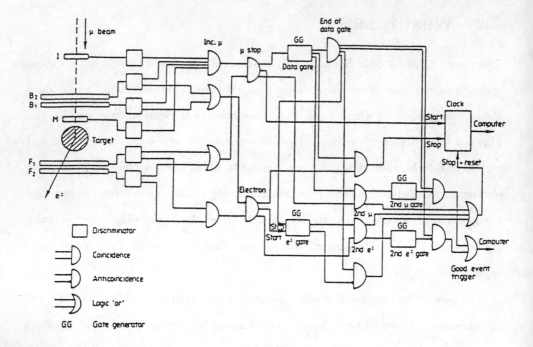

Figure 2.1: A section of the blueprint from the 1990-1991 proposed Brookhaven National Laboratory muon beam line. Shown is the production target T, the bending magnet B1, the channel, the wedge bending magnet B2, and the sample target TGT. Not shown is a separator/spin rotator, which was not a part of this beam line but is a part of most muon beam lines.

2.3   Pion and Muon Production

Most muons which occur naturally are the result of the decay of a pion which has been produced due to the high energy nuclear collision of a cosmic ray with a particle in the earth's upper atmosphere. These muons shower the earth, but their flux density is too low for experimentation. Therefore muons are most often produced by re-creating the cosmic ray collision in a laboratory. The following reactions between proton beams and target nuclear protons result in pions:

$$p + p \rightarrow p + n + \pi^+$$

$$p + n \rightarrow n + n + \pi^+$$

$$\rightarrow p + p + \pi^-$$

Pions can also be produced with high energy electrons. Upon hitting the target, the electrons produce other electrons, positrons, and photons which form a shower of particles. Photons of the right energy ($\geq$ 300 MeV) can interact with nuclei in the target and produce pions. Beams have been produced by this means at both Saclay and NIKHEF, and the possibility of such a muon beam line at CEBAF is discussed in a report by Kossler [9].

The lifetime of the charged pion is approximately 26 ns, at which time it spontaneously decays via:

$$\pi^+ \rightarrow \mu^+ + \nu_\mu$$

where ν_μ is the neutrino associated with the muon and enters to conserve lepton number. The pion has zero spin, therefore conservation of spin angular momentum insists that the $\mu^+ - \nu_\mu$ pair must also be zero spin. It is known [10] that neutrinos are spin $\frac{1}{2}$, left-handed particles – $i.e.$ their spin is anti-aligned with their momentum, or their helicity is -1. The spin of the muon after pion decay must therefore point opposite to its momentum (in the

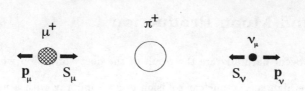

Figure 2.2: A schematic of positive pion decay into a positive muon and its associated neutrino (in the rest frame of the pion). Momenta **p** and spin directions **S** are labeled.

pion rest frame). See Fig. 2.2. This longitudinal polarization of the muon is crucial to μSR, as will become clear below.

Clearly, the muon's energy, momentum, and polarization direction in the LAB are going to depend on how and where it was created. If the pions with little or no kinetic energy near the surface of the target decay, then the muons which decay into the beam line have relatively low energy and are nearly 100% longitudinally polarized. This type of beam is called a *surface* beam. The kinematics of the surface muons can be worked out using the relativistic invariance of mass and energy equation, as well as conservation of momentum. The results are:

$$p_\mu = \frac{(m_\pi^2 - m_\mu^2)c}{2m_\pi}$$
$$E_\mu = \frac{(m_\pi^2 + m_\mu^2)c^2}{2m_\pi}$$
$$T_\mu = \frac{(m_\pi^2 - m_\mu^2)c^2}{2m_\pi}$$

where p_μ, E_μ, and T_μ are the muon's momentum, total energy, and kinetic energy, respectively, and m_π and m_μ are the pion and muon masses, respectively. Using the following values for the masses[11]:

$$m_\pi = 139.57 MeV/c^2$$

$$m_\mu = 105.67 MeV/c^2$$

one finds:

$$p_\mu = 29.79 MeV/c$$

$$E_\mu = 109.78 MeV$$

$$T_\mu = 4.12 MeV$$

The low kinetic energy of the surface muons means that their range in materials is relatively short – $\sim$ 150 mg/cm^2 in air[8]. This is both a problem and a blessing. It is a problem in that it is more difficult to deliver the muon to the target through windows and scintillators – these must be relatively thin as seen by the muon. However, the blessing is that, when constructed properly, the surface beam allows many more muons to stop within a less massive target than with a decay beam. Surface beams are therefore the line of choice for those doing muonic studies of liquids and gases, as well as those studying smaller, less massive solid samples.

A *decay* beam, by contrast, consists of muons which are produced by those pions which exit the target with higher energy. The momentum most often selected is that corresponding to *backward* muons – those which come out of the decay in a direction opposite to that of the parental pion's momentum. The kinematics of these muons can also be worked out using the same conservation laws as above, except now $\mathbf{p}_\pi = \mathbf{p}_\mu + \mathbf{p}_\nu$. The muon's momentum is:

$$p_\mu = \frac{(m_\pi^2 - m_\mu^2)(p_\pi^2 + m_\pi^2 c^2)^{1/2} \pm p_\pi(m_\pi^2 + m_\mu^2)}{2m_\pi^2}$$

where $(+)$ is for forward and $(-)$ is for backward muons. Taking a representative value for p_π of 200 MeV/c[6], which corresponds to a kinetic energy of 104.7 MeV, and putting in the numbers used above yields:

$$p_\mu^f \;=\; 209.37\,MeV/c$$

$$T_\mu^f \;=\; 128.66\,MeV$$

$$p_\mu^b \;=\; 105.25\,MeV/c$$

$$T_\mu^b \;=\; 43.77\,MeV$$

where superscript f means forward and superscript b means backward. Due to the high energy and momentum of the forward muons, the backward muons are preferred. It should also be noted that the polarization of a backward muon, which in the pion rest frame is opposite to its momentum, is parallel to its momentum in the LAB frame.

2.4 Muon Decay

The decay of muons is more complicated because it is a three-body decay:

$$\mu^+ \rightarrow e^+ + \nu_e + \bar{\nu}_\mu$$

where ν_e is the neutrino associated with electrons and $\bar{\nu}_\mu$ is the anti-neutrino associated with the muon. The energetics for this decay for the case where both neutrinos go off opposite to the positron are found to be:

$$p_{e^+} \;=\; \frac{(m_\mu^2 - m_e^2)c}{2m_\mu}$$

$$E_{e^+} \;=\; \frac{(m_\mu^2 + m_e^2)c^2}{2m_\mu}$$

$$T_{e^+} \;=\; \frac{1}{2}m_\mu c^2 - m_e c^2 \left(1 - \frac{m_e}{2m\mu}\right) = 52.32\,MeV$$

On average the positron energy is closer to $35\,MeV$, due to the distribution in angle between decay neutrinos.

The direction of the emitted positron with respect to the muon's polarization can be understood in detail by calculating the decay probability for a positron of energy between

$\epsilon \to \epsilon + d\epsilon$ into an angle between $\theta \to \theta + d\theta$. This has been done [12, 13], and the result is:

$$dW^{\pm}(\epsilon, \theta) = \frac{G^2 m_\mu^5 (3 - 2\epsilon)[1 \mp (1 - 2\epsilon)/(3 - 2\epsilon) \cos \theta] \epsilon^2}{192 \pi^2} \, d\epsilon \, d(\cos \theta) \qquad (2.1)$$

where $(-)$ is for μ^-, $(+)$ is for μ^+, θ is the angle between the muon polarization and the momentum of the outgoing positron, ϵ is the reduced energy E_μ/E_{max}, and $G = 1.166 \times 10^{-5} GeV^{-2}$ is the Fermi coupling constant. The important aspect of this equation for our discussion is the $\cos \theta$ term. It states that the distribution in angle of the emitted positron is a $1 \mp$ function of energy multiplying $\cos \theta$. Therefore the direction of the emitted positron for positive muons is mainly forward, and for the maximum energy case it is exactly forward. This can be understood qualitatively from looking at the maximum energy situation as depicted in Fig. 2.3. The neutrinos' spin directions cancel by virtue of their being a particle–antiparticle pair. This means that the positron spin direction must be the same as that of the muon. Case a shows the positron coming off with momentum opposite to the muon spin direction; hence it has negative helicity. Case b shows the positron coming off with its momentum parallel to that of the muon spin, and hence it has positive helicity. If parity were conserved, then both types of decay would be equally likely. However, case a is not observed in nature; therefore parity is not conserved in this decay and all of the positrons come off with positive helicity in directions which are mainly forward, as stated above. For this case the angle θ is zero and $\epsilon = 1$, and the $\cos \theta$ term in equation 2.1 is 2.

2.5 Thermalization

It was mentioned above that knowing the muon polarization is of crucial importance. First, we have been able to understand the origin of the muon polarization from pion decay, and that its direction parallel to the muon momentum is largely maintained throughout

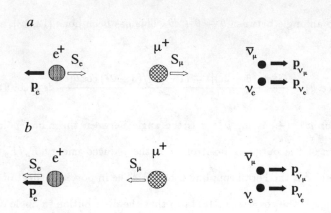

Figure 2.3: Muon decay in the muon rest frame. This situation depicts the maximum positron energy case, where the positron and the neutrinos go off in opposite directions. Case *a* is not observed in nature, while case *b* is.

the beamline and up to the target. Second, we know that *after stopping*, the muon decays into a positron which exits (to first order) in the direction of the muon's spin at the time of decay. It therefore remains to discuss the effects of *thermalization* – the stopping of the muon in the target – on the muon polarization. First thermalization will be discussed, and then its effects on the polarization will be described.

One can imagine that a muon impinging on a material would get knocked about ferociously and lose any sense of order it previously had. However, it must be remembered that we are dealing with point particles and that their interactions, which involve electrostatic forces only (spin independent), occur at very small time scales over small distances. The traditional view of thermalization is similar to the following[7, 6, 14]. Initially the muon loses energy via ionizing atoms and scattering off of electrons. This takes place until the muon has roughly 35 keV or less of kinetic energy and takes a time of the order $10^{-10} - 10^{-9}s$. At this point it is believed that muonium forms – an atom consisting of one muon and one electron – and breaks and reforms until the energy of the muon is about 100 eV. This step requires roughly 10^{-13} s, and the velocity of the muon is roughly

0.002c. Lastly, the muonium becomes relatively stable, losing energy by colliding elasti-
cally with host atoms and inelastically with phonons down to energies of about 15 eV. At
this point one of two things happens. If the material is a semi-conductor or an insulator
the muon will tend to become a part of the molecular make-up of the system with its
electron chemically bound with others[14] – *i.e.* it remains as muonium. If the material
is a metal, then the muon becomes free from the muonium electron and is screened by
the free charges. (In both cases one should note that when the positive muon/muonium
stops, it stops *interstitially* in the material.)

The polarization of the muon *is* affected during all of this, but the effects are negligible.
It was mentioned above that during the initial phases of stopping there is scattering from
electrons. After calculating cross sections for scattering of longitudinally polarized muons
from unpolarized electrons, Ford and Mullin [15] found that the depolarization of the
muons is proportional to the fractional energy loss due to the scattering:

$$Depol \propto U = \frac{m_e}{m_\mu}\beta^2 \sin^2\frac{\alpha}{2}$$

where U is the fractional energy loss, $\beta = v/c$, and α is the center of momentum scattering
angle. The key term is $\beta^2 m_e/m_\mu$, which is very small and gets smaller as the muon slows.
Therefore electron scattering has small depolarization effects on the muon.

The middle stages of thermalization are the muonium stages, where the muon is cycling
through periods of being bound and free and its speed is similar to the orbital speed of
valence electrons. The amount of depolarization occuring here depends on how long the
muon is in a state of muonium, which in turn depends heavily on the density of the
material. That is, denser materials like metals cause more collisions, and hence more
cycles of bound/free for the muon, in a shorter time. Therefore the amount of time
the muon is bound in muonium, which primarily causes depolarization, is less in dense
materials and the net loss of depolarization is less. The depolarization in liquids and
gasses is significantly higher [7] due to their lower density.

The last and lowest energy stage also causes little depolarization in metals. Once again the time period is extremely short, and the conduction electrons present quickly cause any bound muons to become "free," *i.e.* shielded, leaving the muon with a polarization vector pointing in essentially the same direction as it started at higher energy.

Chapter 3

The μSR Technique

This chapter describes the μSR techniques which are employed in most experiments to-day. Transverse, longitudinal, and zero field geometries are discussed and the kinds of experimental results often obtained are shown. A brief introduction to precession is given and applied to the time evolution of the muon's polarization.

3.1 Time Differential μSR

Time differential μSR (TD-μSR) is the experimental procedure most often employed by experimenters today. This procedure is described in reference to Fig 3.1. A typical TD-μSR experimental arrangement consists of: Helmholtz coils, plastic scintillators, a target, and the associated vacuum and cryogenic equipment. The elements are labeled in the figure, but the vacuum and cryogenic equipment are left out for clarity.

In the figure, muons enter one at a time from the left and traverse the scope scintillator S. The muon then encounters the target m and either stops (thermalizes) or continues out the back of the target to the veto scintillator V. The scope and veto are connected to a coincidence box which will reject any events corresponding to a muon not stopping. If the muon does stop, the scope pulse becomes a start signal for a time-to-digital converter

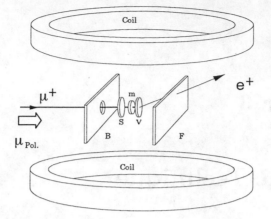

Figure 3.1: A typical experimental setup for μSR experiments. Muons enter at left and encounter the scope detector S, the target m, and possibly the veto counter V. B and F are backward and forward positron detectors, and the coils produce an external magnetic field.

(TDC). The TDC stop signal is obtained when one of the large plastic scintillators (B,F) around the sample detects the decay positron from muon decay. Typically the TDC signal is stored in a buffer memory and then periodically transferred to a computer in which histograms are later generated and used for analysis.

A typical schematic of a μSR detection and electronics system is shown in Fig. 3.2. The electronics is configured to take care of certain contingencies which may arise. For example, the TDC will accept a stop signal up until a certain time has elapsed (a time *window* usually about $5\tau_\mu \sim 10\mu s$). If there is no positron after this time the whole event is thrown out and things start over.

Other possibilities include two muons stopping before the window has expired. There are two cases for this. In one, the second muon arrives before the first decays – an obvious problem in distinguishing which positron comes from which muon. In the other, the second muon comes after the first decays, but still within the time window. For this

17

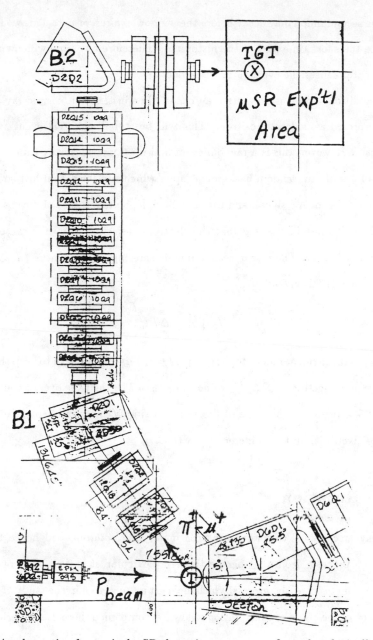

Figure 3.2: A schematic of a typical μSR detection system configured to logically reject bad events as discussed in the text. Taken from Ref. [6].

case the second muon may decay before the window expires, or not. This would artificially enhance the short time part of the histogram. Therefore any time window in which two muons enter causes the event to be discarded.

In addition, accidental events may occur. For instance, a cosmic ray (muon) may be interpreted as a decay positron. This will be stored as a real event as part of the histogram. However, this is a random event and may happen at any time. It is therefore a part of a uniform, random background upon which sits the actual histogram. It is like a DC offset for an AC signal, and can, in principle, be taken care of during analysis.

Finally, it must be realized that the electronic rejection of events causes the overall count rate to suffer. The rate of second muons entering the apparatus $I_{second\mu}$ during the time window ΔT is given by[7]:

$$I_{second\mu} = \Delta T I_\mu^2 \tag{3.1}$$

where I_μ is the incident muon stop rate. If $\Delta T = 10\ \mu s$ and we wish to keep the percentage of two muon rejections $I_{second\mu}/I_\mu$ to no more than 5%, then the overall count rate upper limit is $5 \times 10^3\ s^{-1}$. This type of limitation is simply an experimental fact of life, and must be lived with in time differential μSR.

3.2   Rotation

As mentioned above, this work concerns muon spin *rotation*, and hence the following discussion proceeds from that perspective. In particular, it is the ability of the muon to rotate, or, more correctly, *precess*, which makes it a useful tool in the study of materials. A discussion of the origin of this rotation follows, accompanied thereafter by brief discussions of various μSR techniques.

Of the many characteristics of muons, those important for understanding rotation are *spin* and *magnetic moment*. Quantitatively, the magnetic moment of the muon $\mu_\mu =$

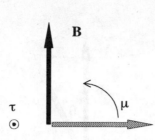

Figure 3.3: The torque produced on a magnetic moment when experiencing a magnetic field **B**.

1.001165932$\frac{e\hbar}{2m_\mu c}$ [11], and its spin $S_\mu = \frac{1}{2}$. Qualitatively, the knowledge of 1/2 spin allows the use of a vector picture to describe the muon's behavior. Any magnetic moment, be it a particle or a bar magnet, will experience a torque when placed in a magnetic field. The torque will act to orient the magnetic moment vector so that it is parallel with the field – Fig. 3.3. Mathematically, the situation is described by:

$$\tau = \mu \times \mathbf{B}$$

If the moment is spinning, then the torque acts so as to cause the magnetic moment vector to precess, or rotate, about the direction of the magnetic field. This is depicted in Fig. 3.4. We know from classical mechanics that the torque equals the time rate-of-change of the angular momentum, which in this case is spin angular momentum: $\tau = d\mathbf{S}/dt$. It is also true that the magnetic moment is proportional to the spin via:

$$\mu = \gamma \mathbf{S}$$

where γ is the gyromagnetic ratio of the spin under consideration. Putting this together

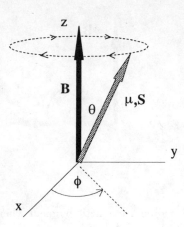

Figure 3.4: The torque on a spinning magnetic moment causes it to precess about the direction of the magnetic field.

yields:

$$\frac{d\mathbf{S}}{dt} = \gamma \mathbf{S} \times \mathbf{B}$$

This equation defines the direction of precession of the moment about the field, as seen in Fig. 3.4. From geometry we see that $|d\mathbf{S}| = S\sin\theta d\phi$, and using $|\mathbf{a} \times \mathbf{b}| = ab\sin\alpha$ we get:

$$S\sin\theta\frac{d\phi}{dt} = \gamma S B \sin\theta$$

which leaves:

$$\frac{d\phi}{dt} = \gamma B = \omega$$

This is the frequency of precession of the moment about the field. The linear relation between field and precession frequency is of vital importance and forms the basis for the use of muons as probes of magnetic field distributions.

3.3   Time Evolution of the Muon Polarization

In order to describe the use of muons as probes of magnetic field distributions we need to develop the time evolution of the polarization vector. This is done by finding its components as functions of time along three mutually perpendicular directions. The results of this development will aid in the description of the various μSR experimental techiques in the following sections.

We first consider a muon stopped in some material with its polarization pointing in a general direction – Fig.3.5. There exist various definable directions. One is the average field **B** direction, which we also define as the z direction. The second is a crystalline direction of the material, termed **c**. **c** is defined to lie within the $x - z$ plane with polar angle θ, thereby determining the coordinate system axes. There exists a local field **b** at the muon site which in general is *not* parallel to the average field **B**. The direction of **b** is given by the polar angle δ and azimuthal angle ϕ_b. Lastly, the polarization is initially at a polar angle α and azimuthal angle ϕ_p.

To find the time evolution of the muon's polarization, we break the polarization vector into two components, one parallel to **b** and one perpendicular to **b**. The one parallel to **b** is independent of time and is found by the dot product $\mathbf{P}(0) \cdot \hat{\mathbf{b}}$. The perpendicular component is obtained from $\mathbf{P}(0) \times \hat{\mathbf{b}}$ and $\mathbf{P}(0) - \mathbf{P}(0) \cdot \hat{\mathbf{b}}$, and rotates at an angular frequency $\omega = \gamma b$. These two components can then be projected onto any direction. A compact representation for the polarization as a function of time is via the following equation:

$$\mathbf{P}(t) = \left(\frac{\mathbf{P}(0) \cdot \mathbf{b}}{b}\right) \frac{\mathbf{b}}{b} + \left[\mathbf{P}(0) - \left(\frac{\mathbf{P}(0) \cdot \mathbf{b}}{b}\right) \frac{\mathbf{b}}{b}\right] \cos \omega t + \left(\frac{\mathbf{P}(0) \times \mathbf{b}}{b}\right) \sin \omega t \qquad (3.2)$$

To find a component of the polarization one simply dots this expression into a direction of interest.

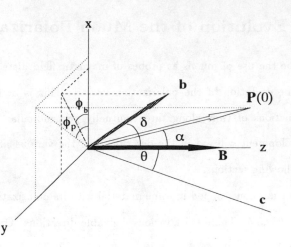

Figure 3.5: General orientations for the muon initial polarization $\mathbf{P}(0)$, the average field $\mathbf{B}$, the local field $\mathbf{b}$, and the crystal direction $\mathbf{c}$.

3.4　The Transverse Field (TF) μSR Technique

The most popular TD-μSR technique is the transverse field technique. The experimental arrangement in this case is one where the muon's initial polarization direction is perpendicular to an externally applied magnetic field. This situation is depicted in Fig 3.1, with the muon's polarization in the plane perpendicular to the field applied by the coils.

As discussed above, the muon will precess in this plane (neglecting local effects) until it decays. The probability of the muon still being around at a time t is $n(t) = e^{-t/\tau_\mu}$. From equation 2.1 we see that the muonic decay as a function of angle is of the form $1 + a \cos \theta$. If now we let the muon precess at a frequency ω we can rewrite this equation as $1 + A_o P(t)$, where A_o is an initial asymmetry and $P(t)$ is from equation 3.2. The precession will cause modulations in the decay curve, yielding the expression:

$$n_d(t) = n_o e^{-t/\tau_\mu}[1 + A_o \mathbf{P}(t) \cdot \hat{\mathbf{d}}] \tag{3.3}$$

where the term $\mathbf{P}(t) \cdot \hat{\mathbf{d}}$ is now the projection upon a detector direction $\hat{\mathbf{d}}$. In experiments the muons stop at different sites within the target, and in general will sample many local fields **b**. The term $\mathbf{P}(t) \cdot \hat{\mathbf{d}}$, while strictly correct for one muon, will for an experiment correspond to an average over the sample volume, or, equivalently, the field distribution. That is, the net or average polarization vector $< \mathbf{P}(t) >$ will precess about some average field $< B >$ at a frequency $< \omega >= \gamma_\mu < B >$ and have a time dependent amplitude which is found by:

$$g(t) = \int n(\mathbf{B})P(t; \mathbf{B})d\mathbf{B} \qquad (3.4)$$

where $n(\mathbf{B})$ is the magnetic field distribution in three dimensions throughout the volume of the sample. The histogram equation for many muons now looks like the following (with the experimental background term added):

$$N(t) = N_0 e^{-t/\tau_\mu}[1 + g(t)] + B \qquad (3.5)$$

Defining $g(t) \equiv A_o G(t) \cos(< \omega > t + \phi)$ gives the alternative expression:

$$N(t) = N_0 e^{-t/\tau_\mu}[1 + A_o G(t) \cos(< \omega > t + \phi)] + B \qquad (3.6)$$

where ϕ is an initial phase angle which can take into account both polarization misalignment and electronic dead time, and B is the background term. Note here that $G(t)$ for a TF experiment is the envelope of the asymmetry function, the so called relaxation function.

In many materials where there are static fields, the assumption of a gaussian field distribution for the z component at the muon is valid, and for a large enough transverse field the effects of the local fields can be ignored. This leaves a one dimensional integral:

$$g_x(t) = \int n(B_z)P_x(t; B_z)dB_z \qquad (3.7)$$

where now:

$$n(B_z) = \frac{\gamma_\mu}{\sqrt{2\pi}\Delta_z} \exp\left(-\frac{\gamma_\mu^2 B_z^2}{2\Delta_z^2}\right) \qquad (3.8)$$

where Δ_z is the width or second moment of the gaussian distribution, and from equation 3.2: $P_x(t; B_z) = \cos(\gamma_\mu B_z t) = Re\ \exp(i\gamma_\mu B_z t)$. This results in a $G_x(t)$ which looks like:

$$G_x(t) = \exp\left(-\frac{\Delta_z^2 t^2}{2}\right) \tag{3.9}$$

This development demonstrates the important concept that the observed function $g_x(t)$ is related to the magnetic field distribution $n(B_z)$. Since B and ω are essentially the same thing, we see that it is the Fourier Transform which relates the two quantities. As an example it will now be shown that $n(B_z) = FT(g_x(t))$:

$$n(B_z) = \int_{-\infty}^{\infty} e^{-i\omega t} g_x(t) dt = \int_{-\infty}^{\infty} dt \int_{-\infty}^{\infty} e^{-i\omega t} n(B_z') P_x(t; B_z') dB_z' \tag{3.10}$$

plugging in from equation 3.2 for $P_x(t; B_z')$ we get:

$$n(B_z) = \int_{-\infty}^{\infty} dB_z' \int_{-\infty}^{\infty} n(B_z') e^{-i(\omega_z - \omega_z')t} dt \tag{3.11}$$

Recognizing that:

$$\delta(\omega_z - \omega_z') \equiv \int_{-\infty}^{\infty} e^{-i(\omega_z - \omega_z')t} dt \tag{3.12}$$

leaves:

$$n(B_z) = \int_{-\infty}^{\infty} n(B_z') \delta(\omega_z - \omega_z') dB_z' \tag{3.13}$$

which ends the proof.

It is therefore possible with this μSR technique to almost directly measure the field distribution in metals. This has been the fundamental contribution of μSR. All other quantities extracted from μSR data have been found via indirect methods and/or with assumptions about the field distribution. (Later in this work we will develop a technique which will allow the experimenter to measure the field distribution without using a gaussian, or any other, assumption about the field distribution.) Of course, even in the above proof a gaussian field distribution was assumed. This has shown to be a quite valid assumption in metals where there are static dipolar fields at the muon (*i.e.* the TF relaxation function has a gaussian envelope). This assumption breaks down when the muon hops or, equivalently, when there is field motion. For motion there are other theories and analysis techniques. The interested reader can peruse references [7, 16].

Questions have also been raised as to how well the Fourier Transform of $g(t)$ can reproduce the actual field distribution. What does one get when the gaussian assumptions are not made? How do phase angles from possible misalignments of detectors and/or the initial polarization of the muon affect the transforms? These questions have been addressed, and in an effort not to re-invent the wheel the reader is referred to the PhD thesis of Riseman [17].

An actual TF histogram is shown below in Fig. 3.6, where a gaussian-like envelope of the oscillations is evident. Note the early time part of the graph. This is the electronics dead time – the time it takes the logic to decide whether or not it has a muon. This part of the spectrum is rarely displayed, but is often used in determining the random background for each detector.

Knowledge of the background (before fitting) is mainly used in an alternative representation of the data in which paired histogram data (*i.e.* F and B) is combined to allow the direct extraction of g(t):

$$g(t) = \frac{(N_1 - B_1) - \alpha(N_2 - B_2)}{(N_1 - B_1) + \alpha(N_2 - B_2)} \tag{3.14}$$

where N_1 and N_2 correspond to the raw data from each histogram, B_1 and B_2 are the backgrounds, respectively, and α is an experimentally determined parameter meant to correct for physical differences between the detectors (*i.e.* efficiencies and geometries). Data represented via this scheme is shown in Fig. 3.7.

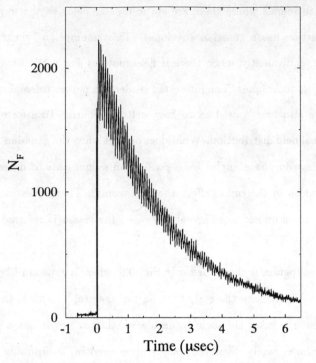

Figure 3.6: A μSR time histogram showing the muonic lifetime of $2.19\mu s$ and a precession signal. Also shown is the "negative time" data used in background calculation.

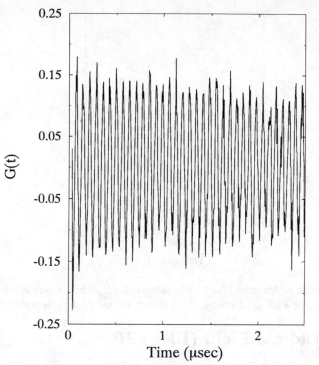

Figure 3.7: Data in the asymmetry representation.

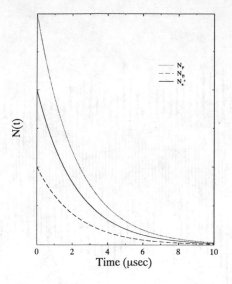

Figure 3.8: Schematic of forward (N_F) and backward (N_B) histogram data for a non-magnetic sample under a LF-μSR study. Also shown is the muonic decay curve.

3.5   Longitudinal Field (LF) μSR

In LF-μSR the muon's initial polarization is parallel to the external field, which is usually much smaller than in a TF-μSR experiment. The positron detectors are generally placed along an axis parallel to the initial muon polarization direction.

The histograms can be represented by:

$$N(t) = N_0 e^{-t/\tau_\mu}[1 + A_o G_z(t)\cos\phi] + B \qquad (3.15)$$

where ϕ is usually $0°$ or $180°$, depending on which detector one is viewing. The shape of $G_z(t)$ will depend on the type of sample under study. If the sample is magnetic, then $G_z(t)$ will generally have some wiggles as a result of muons precessing about fields which are slightly off of the applied field. For non-magnetic samples there will be no precessing, and the histograms will look roughly like Fig. 3.8.

The static magnetic field distribution is now a shifted gaussian[16]:

$$n(B_z) = \frac{\gamma_\mu}{\sqrt{2\pi}\Delta_z} \exp\left(-\frac{\gamma_\mu^2(B_z - B_A)^2}{2\Delta_z^2}\right) \qquad (3.16)$$

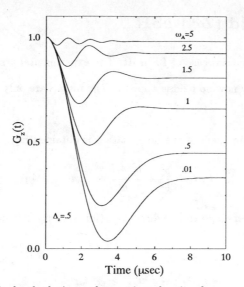

Figure 3.9: Theoretical calculations of equation showing how an increasing applied field causes a decrease in the dephasing of the muons.

which yields the following for a relaxation function[6]:

$$g_z(t) = 1 - \frac{2\Delta_z^2}{\omega_A^2}[1 - e^{-\frac{\Delta_z^2 t^2}{2}}\cos\omega_A t] + \frac{2\Delta_z^4}{\omega_A^3}\int_0^t e^{-\frac{\Delta_z^2 t'^2}{2}}\sin\omega_A t' dt' \qquad (3.17)$$

where $\omega_A = \gamma_\mu B_A$ is the applied field.

The shape of this function depends on the strength of the external field. For small applied fields (ω_A) a random distribution of fields causes each muon to precess at different frequencies with differently directed precession cones. Therefore although all muons start in phase, they quickly become dephased and the value of $G_z(t)$ decreases. If ω_A is quite a bit larger than the values of the local fields, then the frequencies and precession cones are all about the same and a much smaller amount of dephasing occurs. This field dependence is shown in Fig. 3.9.

Once again there are effects due to the motion of either the muons or the fields at the muon sites. The interested reader can read references [7, 16].

3.6   Zero Field (ZF) μSR

ZF-μSR is an obvious special case of LF-μSR. The experimental arrangement is exactly the same as before, only now no field is applied. The muons are only subject to the fields intrinsic to the sample.

The longitudinal component of the polarization is obtained from equation 3.2:

$$P_z(t) = \frac{b_z^2}{b^2} + \frac{b_x^2 + b_y^2}{b^2} \cos \omega t \tag{3.18}$$

which when averaged over all possible **b** directions (for isotropic field distribution functions in all directions) yields:

$$< P_z(t) > = \frac{1}{3} + \frac{2}{3} < \cos \omega t > \tag{3.19}$$

If again a gaussian field distribution is assumed then we obtain for the relaxation function:

$$G_z(t) = \frac{1}{3} + \frac{2}{3}(1 - \Delta_z^2 t^2)e^{-\frac{t^2 \Delta_z^2}{2}} \tag{3.20}$$

which is the famous Kubo-Toyabe formula [18].

An interesting aspect of this equation is that at long times it approaches a value of $\frac{1}{3}$, which is a keynote or signature of static gaussian field distributions and shown in Fig. 3.10. It should be noted that this $\frac{1}{3}$ recovery has come into question for some materials under certain conditions where longer time oscillations have been calculated [19, 20].

The usual sources can be consulted if the reader is interested in the motional aspects of this technique.

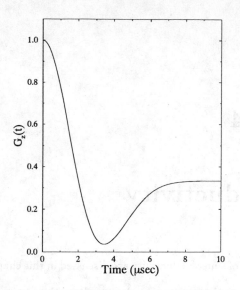

Figure 3.10: A plot of the static Kubo-Toyabe relaxation function. Note the long time recovery to $\frac{1}{3}$.

Chapter 4

Superconductivity

Some of the basic ideas of superconductivity are discussed in this chapter. The penetration depth and coherence length are introduced, as well as the ideas of type I and type II superconductors and their corresponding phase diagrams. The high-T_c superconductor YBCO is shown as a particular example of a type II material. The isotropic London theory is then introduced as a starting point for calculating the microscopic fields within the mixed state of the type II materials. This theory is then extended to anisotropic materials, where a prescription is given for calculating the microscopic field components at points within the equilibrium, anisotropic Flux Line Lattice (FLL). Lastly, the assumptions and limitations of the London theory are addressed.

4.1   Introduction

The first superconductor was discovered in 1911 by a Dutch physicist named Kammerlingh Onnes [21] (actually his graduate student) just three years after he became the first person to liquify helium. The effect Onnes found was that of zero resistance to current flow, which various elements displayed, each at their own characteristic critical temperature T_c. In all, 26 elements (and 11 others under special conditions)[22] as well as a whole host of inorganic and organic compounds are superconductors.

Later, in 1933, another effect of superconductivity was discovered by Meissner and Ochsenfeld [23] – the Meissner effect. A superconductor cooled in a magnetic field will expel the field as it is cooled below its critical temperature. This leaves the material with zero field inside except for in a small layer near the surface where superconducting currents flow. That is, the superconductor becomes a perfect diamagnet, and the applied field lines must bend around the sample. This state exists until either the sample warms above T_c or the field increases above a critical value.

A third characteristic of superconductors is the existence, at least in some form, of an energy gap – Δ. An energy gap for superconductors manifests itself as $\frac{1}{2}$ of the energy needed to break a pair of particles which have condensed into a superconducting state. The existence of a gap means that only energies above twice the gap value are sufficient to break the pair, and that there should be an absence of low energy excitations within a pure material. Most elemental superconductors exhibit behavior consistent with an energy gap which is *isotropic*, or *s-wave*, which means that the surface in momentum space is spherical, and there exists a gap in all directions between the valence and conduction bands. Recently, much research using various techniques (including μSR) has been done in an attempt to probe the energy gap of new superconducting materials, such as $YBa_2Cu_3O_{7-b}$ and $La_{1.85}Sr_{0.15}CuO_4$ and its related compounds. (The first compound, termed YBCO, will be the subject of much discussion in this work). It was initially generally held that they exhibited s-wave properties[24, 25], but then a *d-wave* picture arose [26]. This theory allows for nodes in the gap in certain directions – *i.e.* directions where pairs can be broken with no energy. At present there is no clear, unambiguous body of experimental evidence supporting either model. The following references, on both sides of the argument, are cited for the interested reader[27, 28, 29, 30]. A good theoretical development of the origin of some pairing states is given in [31] and the references cited therein.

A monumental achievement on the theoretical side came in 1957, when Bardeen, Schrieffer, and Cooper published their Nobel Prize-winning work on the microscopic theory of superconductivity[32]. This was the first successful, non-phenomenological theory of superconductivity. It describes how an attraction between electrons can arise within a background of free electrons, allowing them to pair up into Cooper pairs, and form a part of a ground state wave function in which all electrons may be bound. This theory is quite robust, and is still used as a "yard stick" by which other theories and experimental results are measured. Due to its robustness, as well as its complexity, the details of the BCS theory are beyond the scope of this work. The interested reader can consult the original work[32], but will probably have better luck with other people's interpretations[33, 34].

In 1986, researchers at IBM's Zürich laboratory discovered a material which became superconducting at a T_c of $\sim$ 30-35 K[1]. This discovery shook the scientific community because up until then the highest T_c had been only 23.3 K. Soon after, researchers at the Universities of Alabama and Houston developed the material $YBa_2Cu_3O_7$ (called YBCO) which exhibited the even higher critical temperature of $\sim$ 92 K[35], which above the 77 K boiling point of liquid nitrogen and therefore can be made to go superconducting more cheaply than those materials with cooler T_c's. Soon after, even more materials were discovered with even higher critical temperatures and even more interesting characteristics.

These materials caused a lot of excitement in both the academic and industrial research communities throughout the world. It is obvious that there is great scientific and technological potential in the superconductors, and that they deserve a great deal of careful study in order to be understood. A better understanding will not only increase our overall knowledge of our world, but will allow for the best possible applications of the materials that we now have, and will help us to develop even more materials with more useful properties. It is for these reasons that works such as this one are undertaken.

4.2 Fundamental Parameters and Superconductor Type

There are two fundamental parameters associated with all superconductors: λ and ξ. Lambda is the *penetration depth*, the length in the exponential sense in which the magnetic fields die off within a superconductor. ξ is the *coherence length*. It is a fundamental length scale which originated with Pippard[36] and concerns the distance over which local fields within a material have an appreciable effect on the current at a nearby point. The coherence length is often mentioned as a characteristic pair wave packet minimum length, and it is also the order of the size of the flux core in type II superconductors (discussed below). The type of a superconductor is determined by the parameter κ, the Ginzburg-Landau parameter, defined as $\kappa \equiv \lambda/\xi$. Type I superconductors have $\kappa < 1/\sqrt{2}$, while type II superconductors have $\kappa > 1/\sqrt{2}$.

Type I superconductors are the classic superconductors – Pb, Sn, Al, *etc.* Their behavior can be described via Fig. 4.1. For low magnetic fields at a temperature below T_c, $B = 0$, and the curve is linear with a slope of 1. Then at a critical field H_c there is a sharp transition to the normal state. Therefore the magnetic field is either completely expelled or completely allowed within the type I material.

For Type II materials the behavior is as depicted in Fig. 4.2. Here the field is completely expelled up to H_{c1}, but then is allowed to penetrate more and more until an upper critical field H_{c2} is reached, where the field penetrates completely. The region between H_{c1} and H_{c2} is termed the *mixed state* and will be the subject of much discussion in this work.

The mixed state is characterized by magnetic flux entering the samples in units of the flux quantum $\Phi_o = hc/2e = 2.07 \times 10^{-7} G cm^2$. The resulting equilibrium geometry formed is usually that of a triangular lattice, with a flux quantum at each corner of the triangle.

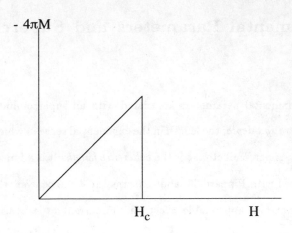

Figure 4.1: A sketch of the magnetic field behavior of a type I superconductor. During the linear part of the graph there is no **B**, so **H** = −4π**M**. At the critical field H_c the superconductor can no longer expel the field, so it becomes completely normal.

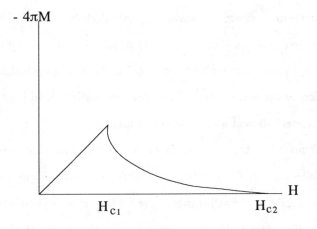

Figure 4.2: A sketch of the magnetic field behavior of a type II superconductor. For fields up to H_{c1} the behavior is like a type I. For larger fields the material enters into the mixed state where flux can enter in the form of filaments of the quantum Φ_o, and a flux line lattice is formed. The material is not completely normal until the value H_{c2} is reached.

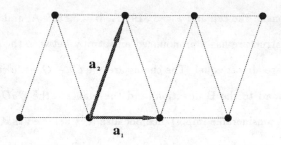

Figure 4.3: A sketch of the equilibrium magnetic flux line lattice for anisotropic, unaxial superconductors. The real space basis vectors a_1 and a_2 are shown.

See Fig. 4.3. It has been shown that the triangular flux line latice (FLL) geometry is most often the proper one for minimizing the free energy of the system at higher and intermediate fields [37]. The question of very low fields will be discussed later.

The density of the FLL is a function of the magnetic induction, going as $n = \frac{B}{\Phi_o} \ cm^{-2}$. When the density gets very high the flux cores begin to overlap. This costs energy, so the sample becomes normal conducting.

4.3 The $YBa_2Cu_3O_{7-\delta}$ Superconductor

An example of a type II superconductor is the relatively new material $YBa_2Cu_3O_{7-\delta}$, termed YBCO for short. This superconductor was discovered in 1989 by T.C. Chu *et al.*[38] and has hence been possibly the most studied of the high-T_c materials. The term high-T_c refers to the then incredibly high critical temperature that this material exhibited of $T_c = 92 \ K$, which is about an order of magnitude higher than most type I superconductors. For reference a sketch of the unit cell crystal structure is shown in Fig.

4.4. The dimensions of the cell are: $\mathbf{a} = 3.8198$ Å, $\mathbf{b} = 3.8849$ Å, and $\mathbf{c} = 11.6762$ Å[39];

hence the crystal structure has a pronounced anisotropy between the $\mathbf{c}$ axis and the $\mathbf{a}$-$\mathbf{b}$

directions, which are almost equal. The chains are the $Cu - O - Cu$ rungs labeled in the

figure and correspond to the $\mathbf{B}$ direction, and the planes – the CuO_2 planes labeled in

the figure – are of considerable importance because it is believed that superconductivity

arises here. The superconducting currents flow most easily parallel to, and with the most

difficulty perpendicular to, the planes. Therefore there is not only a crystalline anisotropy

to these materials, but a superconducting one as well. This superconducting anisotropy

will be addressed below when a theoretical description of anisotropic superconductors

is developed. The superconducting charge carriers, in this case holes, are believed to

originate in the planes. The crystalline structure and electrical behavior of the planes

show almost no anisotropy within the plane. It is for these reasons that the material is

considered uniaxial.

YBCO is the material chosen for study in this work because its properties and behavior

seem to allow a theoretical description within the anisotropic theory just mentioned. Its

properties are well known because of the wealth of information and knowledge obtained

from intense study for many years. These properties will be discussed later in section 4.9.

First, however, we need to develop more on superconductivity.

4.4 Isotropic London Theory

It was mentioned above that the BCS theory was the first successful rigorous theory

of superconductivity. By contrast, the London theory of isotropic superconductors is a

phenomenological theory which is much easier to understand, and it provides a simple

method for the calculation of the magnetic fields of the FLL.

The theory starts by considering the free energy associated with an isolated vortex

within a superconductor in the mixed state. The vortex has a core of radius ξ, and

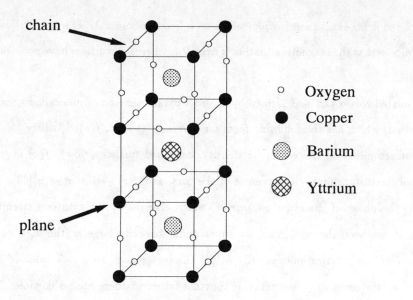

Figure 4.4: A sketch of the unit cell of YBCO.

the associated fields extend out isotropically over a length of order λ. The term "isolated vortex" implies that the magnetic situation is near H_{c1}, where few vortices have penetrated the sample.

The isolated vortex can be described by an order parameter (sometimes called a vortex wavefunction) which has the following form for a vortex: $\psi = \psi_\infty f(r)e^{i\theta}$, where $\mid \psi \mid^2 = n_s$ (the superconducting charge carrier density, described further below), $f(r)$ is some function of the distance r from the center of the core, and $f(r) \to 1$ as $r \to \infty$[33]. The angle θ is the phase of the order parameter, which changes by 2π with one circuit of the core. It has been shown[33] that for superconductors with large κ (the high *kappa* limit) the function f varies quite rapidly within the core region, rises to a value of 1 at a distance on the order of ξ, and remains constant wherever there are no vortices. The order parameter therefore follows this behavior, too. This is significant in that if the wavefunction is constant, then the superconducting charge carrier density is also constant over these regions. If this density is not a constant, then a theory which allows for its variation must be used, such as the Ginzburg-Landau theory[40]. If the superconducting charge carrier density is constant, then one can use the London theory[41], which is discussed next below.

One begins the London theory by writing down the free energy density of the superconductor. There will be a term for the energy density of the magnetic field, given by $b^2/8\pi$. There will also be a term for the kinetic energy density of the superconducting particles. This term is $\frac{1}{2}Mv^2 n_s$, where M is the effective mass of the superconducting particle, v is its drift velocity, and n_s is the superconducting charge carrier density (cm^{-3}). It is important to note again that the charge carrier density is assumed constant, except for right at the vortex core, where it is zero – if this is not true then the London theory will not hold.

We now use the equation for current density:

$$J = n_s e v \tag{4.1}$$

where e is the charge per particle. This yields for the kinetic energy density:

$$E_k = \frac{1}{2}\frac{M}{n_s e^2} J^2 \tag{4.2}$$

Next recall the Maxwell equation:

$$\nabla \times \mathbf{b} = \frac{4\pi}{c}\mathbf{J} \tag{4.3}$$

where c is the speed of light (Gaussian units). Doing some algebra yields:

$$E_k = \frac{1}{8\pi}\frac{Mc^2}{4\pi n_s e^2} \mid \nabla \times \mathbf{b} \mid^2 \tag{4.4}$$

The free energy density of the system is now:

$$\mathcal{F} = \int_V \frac{1}{8\pi}\left(b^2 + \frac{Mc^2}{4\pi n_s e^2}|\nabla \times \mathbf{b}|^2\right) \tag{4.5}$$

where the integral is over a volume V. The term $\frac{Mc^2}{4\pi n_s e^2}$ is recognized as the square of the London penetration depth, λ_L^2[41]. The free energy density now takes on the familiar form:

$$\mathcal{F} = \frac{1}{8\pi}\int_V \left(\mathbf{b}^2 + \lambda_L^2 |\nabla \times \mathbf{b}|^2\right) dV \tag{4.6}$$

In order to find field equations from the free energy density we must perform a variational calculation, minimizing the free energy with respect to variations in the magnetic field b. This can be done by first writing the free energy as a function of $b + \delta b$, and then subtracting off the free energy as a function of b. The result is:

$$\delta\mathcal{F} = \frac{1}{4\pi}\int_V \left(\mathbf{b}\cdot\delta\mathbf{b} + \lambda_L^2(\nabla \times \mathbf{b})\cdot(\nabla \times \delta\mathbf{b})\right) dV \tag{4.7}$$

Integrating the second term by parts yields:

$$\frac{1}{4\pi}\int_V \left(\lambda_L^2(\nabla \times \mathbf{b})\cdot(\nabla \times \delta\mathbf{b})\right) dV =$$
$$\frac{\lambda_L^2}{4\pi}(\nabla \times \mathbf{b})\cdot\delta b + \frac{\lambda^2}{4\pi}\int_V (\nabla \times (\nabla \times \mathbf{b}))\cdot\delta b\, dV \tag{4.8}$$

The first term on the right hand side is recognized as the current J dotted into δb. This term therefore integrates to zero far from the core where the currents go to zero. However, within the core there is a problem because the superconducting charge density is no longer constant. It actually goes to zero at the center of the core, causing a singularity. This is taken care of by introducing a delta function "source term" in the equation which multiplies the flux quantum Φ_o. One can now write the equation as:

$$\delta \mathcal{F} = \frac{1}{4\pi} \int_V \left[\mathbf{b} + \lambda_L^2 \left(\nabla \times (\nabla \times \mathbf{b}) \right) - \Phi_o \delta(\mathbf{r} - \mathbf{r}_\nu) \right] \cdot \delta \mathbf{b} \; dV \tag{4.9}$$

The minimum energy condition is satisfied when the integrand is zero; hence, the equation for the fields is:

$$\mathbf{b} + \lambda_L^2 \nabla \times (\nabla \times \mathbf{b}) = \Phi_o \delta(\mathbf{r} - \mathbf{r}_\nu) \tag{4.10}$$

This equation, in principle, allows the calculation of the magnetic field $\mathbf{b}(x, y)$ in the mixed state of an isotropic superconductor. A method of doing so will be described in the next section, where the London theory is extended to the anisotropic, uniaxial superconductors.

4.5   Anisotropic, Uniaxial London Theory

The London theory for anisotropic, uniaxial superconductors starts out with the free energy, also. The development follows exactly the isotropic case until equation 4.6. At this point the uniaxiality needs to be included, and this is done using the charge carrier masses both within the superconducting planes and perpendicular to these planes. The mass within the plane is $M_X = M_Y = M_1$; the mass perpendicular to the plane is $M_Z = M_3$, which is more massive and reflects the difficulty of current flow in this direction. The mass is now a tensor with M_1's and an M_3 on the diagonal as represented in the crystal frame. It is generally more convenient to normalize the masses with respect to an average mass defined by $M_{av} = (M_1^2 M_3)^{1/3}$, which leaves:

$$m_1 = \frac{M_1}{M_{av}} \tag{4.11}$$

$$m_3 = \frac{M_3}{M_{av}} \tag{4.12}$$

If we now go to a frame where the geometry is defined as depicted in Fig. 4.5, the mass tensor needs to be transformed. The angle θ is the angle between the crystal **c** axis and the direction of the average field **B**, the x axis is defined to lie in the plane of **c** and **B**, and the y axis is perpendicular to this plane. The mass tensor can be represented in this frame by transforming the crystal mass tensor via a rotation about the Y (or y) axis, leaving:

$$m_{xx} = m_1 \cos^2 \theta + m_3 \sin^2 \theta \quad m_{xy} = m_{yz} = 0$$
$$m_{zz} = m_1 \sin^2 \theta + m_3 \cos^2 \theta \quad m_{yy} = m_1 \tag{4.13}$$
$$m_{xz} = (m_1 - m_3) \sin \theta \cos \theta$$

The uniaxiality is folded into the London theory via these masses through the penetration depth λ_L^2[42, 43], where:

$$\lambda_L^2 = \frac{Mc^2}{4\pi n_s e^2} \longrightarrow \lambda^2 m_{ij} = \frac{M_{av} c^2}{4\pi n_s e^2} m_{ij} \tag{4.14}$$

where λ is now an effective penetration depth, and the subscripts on m_{ij} are x, y, or z.

It is often convenient when speaking in these terms to introduce an anisotropy parameter $\Gamma \equiv \frac{m_3}{m_1}$, which for YBCO is 25 [44] and for the Bi-2212 compound is ~ 3025 [45]. It should also be noted that knowledge of Γ is sufficient for finding the masses m_3 and m_1: $m_3 = \Gamma^{2/3}$ and $m_1 = \Gamma^{-1/3}$. With these changes the free energy density becomes:

$$\mathcal{F} = \frac{1}{8\pi} \int_V \left[\mathbf{b}^2 + \lambda^2 m_{ij} (\nabla \times \mathbf{b})_i (\nabla \times \mathbf{b})_j \right] dV \tag{4.15}$$

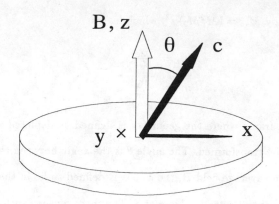

Figure 4.5: Diagram showing the angle θ between the average field $\mathbf{B}$ and the crystal axis **c**.

The field equations are obtained via the same type of minimization derivation used above. The resulting field equations are:

$$b_k = \lambda^2 m_{ij} \epsilon_{ist} \epsilon_{jkl} \frac{\partial^2 b_s}{\partial x_t \, \partial x_l} + \Phi_o \delta_{zk} \delta(\mathbf{r} - \mathbf{r}_\nu) \qquad (4.16)$$

where the equation is written in component form using the summation convention, and ϵ_{abc} is the fully anti-symmetric tensor. This equation, like its isotropic partner equation 4.10 above, allows in principle the calculation of the fields within the superconductor.

4.6 Calculation of Fields in Anisotropic Superconductors

The method described here for calculating the magnetic field at any point within a FLL in an anisotropic, uniaxial superconductor, follows closely the method described by Thiemann, et al. [46]. Similar work has been done by Balatskiĭ and others[47, 48], and more recently this development has been used as a part of the thesis of Riseman[17].

We first expand equation 4.16 and, recognizing the following general vector identity $\nabla^2 \mathbf{A} = \nabla(\nabla \cdot \mathbf{A}) - \nabla \times (\nabla \times \mathbf{A})$, write the field components as:

$$b_x - \lambda^2(m_{zz}\nabla^2_{xy}b_x - m_{xz}\frac{\partial^2 b_z}{\partial y^2}) = 0$$

$$b_y - \lambda^2(m_{zz}\nabla^2_{xy}b_y + m_{xz}\frac{\partial^2 b_z}{\partial x \partial y}) = 0 \qquad (4.17)$$

$$b_z - \lambda^2(m_1\frac{\partial^2 b_z}{\partial x^2} + m_{xx}\frac{\partial^2 b_z}{\partial y^2} - m_{zz}\nabla^2_{xy}b_x) = \Phi_o\sum_\nu \delta(\mathbf{r} - \mathbf{r}_\nu)$$

where $\nabla^2_{xy} = \frac{\partial^2}{\partial x^2} + \frac{\partial^2}{\partial y^2}$ is the two dimensional Laplacian in the xy plane, $\mathbf{r}$ is some vector in the FLL, and $\mathbf{r}_\nu$ are vectors to the vortices.

These equations are strictly correct for an isolated vortex, but can be used with good confidence for situations where there are many vortices which are well spaced and do not overlap. We will assume for the moment that we have vortices which are well separated in the equilibrium arrangement shown in Fig. 4.3. With this assumption, we now invoke the well known periodicity of the FLL, which enables us to perform a Fourier transformation. What we will do is transform the equations into Fourier space (reciprocal lattice space), perform a summation in this space, and transform back into real space for the final result.

The Fourier relationship is written as:

$$\mathbf{b}(\mathbf{r}) = \sum_{\mathbf{G}}\mathbf{b}(\mathbf{G})e^{i\mathbf{G}\cdot\mathbf{r}}$$

$$\mathbf{b}(\mathbf{G}) = \frac{B}{\Phi_o}\int \mathbf{b}(\mathbf{r})e^{-i\mathbf{G}\cdot\mathbf{r}}\,d\mathbf{r} \qquad (4.18)$$

where $\mathbf{G}$ are reciprocal lattice vectors and B is the magnitude of the average field over the real space FLL unit cell. The integral is over the unit cell in real space.

Plugging the upper of equations 4.18 into equations 4.17 and solving gives the components of the field as a function of reciprocal lattice vectors:

$$b_x(\mathbf{G}) = B\lambda^2 m_{xz}G_y^2/d$$

$$b_y(\mathbf{G}) = -B\lambda^2 m_{xz}G_x G_y/d$$

$$b_z(\mathbf{G}) = B(1 + \lambda^2 m_{zz}G^2)/d \qquad (4.19)$$

where the denominator d is given by:

$$d = (1 + \lambda^2 m_1 G_x^2 + \lambda^2 m_{xx} G_y^2)(1 + \lambda^2 m_{zz} G^2) - \lambda^4 m_{xz}^2 G^2 G_y^2 \qquad (4.20)$$

These equations – for a uniaxial, anisotropic superconductor – will lead to the real space microscopic magnetic field at some general point within the FLL. When the real space field is found (described below), it will in general have components which are transverse to z, *i.e.* x and y components. These transverse components are a result of the anisotropy of the material being introduced via the mass tensor, and do *not* arise within the isotropic London theory. These transverse components will be the subject of much discussion in chapter 6.

It now remains only to state the real and reciprocal lattice vectors. From various studies it has been found [42, 49] that the basis vectors of the real space, equilibrium FLL, for directions of B with respect to the material axis of $0° \le \theta \le 70°$, are given by:

$$\mathbf{a_1} = \left(\frac{2\Phi_o}{B}\right)^{1/2} \left(\frac{m_{zz}}{3m_3}\right)^{1/4} \hat{\mathbf{x}}$$

$$\mathbf{a_2} = \frac{a_1}{2}\left[\hat{\mathbf{x}} + \left(\frac{3m_3}{m_{zz}}\right)^{1/2} \hat{\mathbf{y}}\right] \qquad (4.21)$$

The angle between the vectors (η) is given by[46] $\tan \eta = (3m_3/m_{zz})^{1/2}$. These vectors map out a FLL which consists of isosceles triangles, and for $\theta = 0$ the triangles are equilateral (isotropic case).

The reciprocal lattice vectors corresponding to these real space vectors are:

$$G_x = \pi \left(\frac{2B}{\Phi_o}\right)^{1/2} \left(\frac{3m_3}{m_{zz}}\right)^{1/4} n \qquad (4.22)$$

$$G_y = \pi \left(\frac{2B}{\Phi_o}\right)^{1/2} \left(\frac{m_{zz}}{3m_3}\right)^{1/4} (2m - n) \qquad (4.23)$$

where

$$m, n = 0, \pm 1, \pm 2, \ldots$$

These vectors map out a reciprocal FLL which consists of isosceles triangles, much like its real space counterpart.

The prescription for the calculation of the local magnetic field b(x,y) within the FLL is now clear:

1. Choose an angle θ, an anisotropy Γ, and an effective penetration depth λ for a material of interest.

2. Calculate the mass tensor m_{ij} for these values.

3. Calculate the components in the upper of equations 4.19.

4. Sum equation 4.18 over a sufficient number of reciprocal lattice vectors for each component of the field.

5. Repeat steps 1-4 above for many points within a FLL unit cell.

More needs to be said about the phrase "sufficient number" in the fourth item above. First, the sum over equation 4.18 may not converge for all points in the FLL due the G dependence in each of the terms. The question becomes one of how many reciprocal lattice vectors are needed obtain reasonable results. A procedure often used is to cut off the sum at an appropriate number of reciprocal lattice vectors, $G_{max} \sim 1/\xi$. Reciprocal lattice vectors of length greater than this fold in Fourier components which are the result of the logarithmic singularity at the core[33]. For a two dimensional reciprocal lattice there will be two different cutoffs, one for each direction of the lattice, and in general these will be different.

A similar method, and the one used in this work, is to perform an empirical investigation as to the degree of convergence for various numbers of reciprocal lattice vectors. Once a number is found which, when exceeded, causes little change in the field values, this number is used as the cutoff. This method was used for my calculations, and I decided on a value of ± 50 for the integers n and m in equations 4.22 above. The field values at nearly all points within the FLL showed little variation for numbers greater than this, and for lesser numbers it was obvious that not enough were used.

4.7 Limitations of the London Theory

The London theory, like all theories, has limitations. The origin of the limitation is that for the London theory to be valid, the superconducting charge carrier density must be essentially constant throughout the superconductor. As discussed above, within the isolated vortex approximation the vortex wave function increases quickly until it reaches a value of ~ 1 at the core edge. Outside the core, the wave function, and hence the superconducting charge carrier density, varies little[33].

The development assumes one is near the lower critical field H_{c1}, and is strictly only valid there. However, for intermediate field values, $H_{c1} \ll H \ll H_{c2}$, the London theory may also be applied with good confidence for materials which are in the *high-κ limit*, because here again the vortex wave function is still largely a constant, and the cores may be taken care of via the delta functions used above.

For high-κ materials, it is only when in the high field limit, $H \sim H_{c2}$, that the London theory begins to break down. In this situation the flux cores are very close together, and there is a significant variation in the vortex wave function. This is a result of the charge density varying, and hence the basic assumption upon which the London theory is based is no longer valid. In such situations one must abandon the London theory in favor of a theory which is valid near the upper critical field, like the famous Ginzburg-Landau theory[40, 34].

An obvious fault of the London theory is that the field at the flux core is logarithmically divergent at the *point* which is assumed to be the core – see, *e.g.* deGennes [34]. Therefore the effect of finite core size may have to be taken into account. It was mentioned above that for equilibrium FLL's the reciprocal lattice sum can be cut off at $G_{max} \sim 1/\xi$[50], which causes the field within the core to reach a maximum, finite level[51]. This method has relatively minor effects on the overall field values within the FLL and only significantly affects the fields very near (or within) the cores. In addition, there are times where the core size is such that this type of cutoff is very necessary. This can be due to actual large ξ values[52], or possibly due to longitudinal fluctuations along the vortex length enhancing the vortex core contribution[53] to the fields within the FLL.

The isotropic and anisotropic theories developed above have additional assumptions. The use of the Fourier ideas assumes a regular, periodic array of vortices in a triangular lattice which has formed within a single crystal. This is the equilibrium condition as solved by Campbell *et al.*[37]. The high-T_c materials, while certainly being in the high-κ limit discussed above, have historically been full of pinning centers which trap flux and cause significant distortion within the FLL. The distortion often manifests itself in the bending of the flux tubes, causing additional fields which are transverse to the previously defined vortex direction, and generally causing disorder in the FLL.

Single crystals pose problems in that they are often twinned, meaning that while there is a well defined crystal c axis, there is no defined a or b direction due to their being skewed at right angles to each other along various planes parallel to the c axis direction. Early studies on YBCO crystals using Bitter pattern techniques showed that the twin boundaries caused severe pinning of flux tubes, increasing the flux density along them by nearly a factor of 2[54]. However, this technique only samples the vortex behavior at the surface of the sample, and we are more interested in the bulk behavior. The bulk is readily probed with neutron scattering, and studies in various twinned single crystals have been performed. These have shown that the FLL becomes slightly distorted if twin planes are present, but that the structure is essentially a triangular geometry[55, 56]. One must be aware that these results have been obtained with relatively high applied fields, on the order of $8 - 10 \ kG$ in most cases. This is because the FLL spacing goes like $d \sim \sqrt{\Phi_o/B}$, and slow neutrons have wavelengths on the order of 10 Å[57]. For the neutrons to "see" the FLL the field must therefore be in this range. Neutrons are therefore an excellent probe of the FLL within the bulk at these fields, but are incapable of seeing anything at lower fields.

However, recently it has been shown[58, 59] that largely untwinned single crystals can be grown. The FLL within these kinds of samples is expected to be nearly distortion free[55] and has been shown to be so in[53].

4.8   Effects of Anisotropy on the FLL

The first effect of the crystalline uniaxial anisotropy on the FLL is the change with angle θ from an equilateral to an isosceles geometry. This effect is the most obvious and comes directly out of the analysis given in section 4.6. The second effect is the existence of the transverse magnetic fields, which for the isotropic case do not exist. In addition, there are other, more subtle, effects of anisotropy which can cause changes within the FLL. These are briefly described below.

At sufficiently weak fields ($H < 100\ G$) the FLL may undergo a phase transformation from a triangular to what has been termed a *chain* state. The effect was predicted theoretically[60, 61, 62, 63, 47, 64, 65] by realizing that the vortices, which normally repel each other, become attracted to each other for large angles θ in these anisotropic materials. The effect has recently been shown to exist in Bitter pattterns[66, 58], and is caused by the interaction potential between the vortices in the high-T_c materials becomming attractive in the x direction. That is, at low enough inductions and high enough angles, the field component b_z becomes negative over a certain range of distances from the core. Since the interaction potential between vortices is proportional to this component of the field[34, 65], the interaction becomes attractive. (For a good theoretical discussion, see ref. [65].)

Once formed, the spacing between fluxons in the chain direction is independent of field for a certain range of fields[65]. Therefore the FLL geometry only changes in one direction as H is varied. In principle the London theory is still valid in this situation, however a different Fourier analysis is needed to properly describe the situation[17].

In addition, some theoretical studies at very low inductions have recently pointed to other equilibrium lattices which are not triangular[67]. The studies show that the triangular lattice at low inductions and with the average field at an angle to the crystal c axis can become unstable and yield to other structures. These studies were pursued in an attempt to explain the phenomenon of flux melting, where it is believed that the FLL may become disordered due to thermally activated forces within the material, as well as the recent sighting of chains mixed within a normal vortex lattice [66]. The theoretical development of this interesting case has been done[68], and this particular instability within the FLL is found to be a result of the high anisotropy of the material and the geometry of the sample. Instead of the free energy density minimization done above, one must now minimize the Gibbs free energy subject to boundary conditions on B and H. This leads to a mixed state within the mixed state, with some flux at an angle to c and some lying parallel to c. This phenomenon is very interesting, but it lies beyond the scope of this work.

Anisotropy in the high T_c materials seems to be linked to the crystal structure. The more anisotropic the crystal structure, the more anisotropic the magnetic and electrical properties. In materials like BSCCO, which has a large ratio of c/a, the superconducting planes can be many angstroms apart, so that the currents and fields associated with one plane may have little interaction with the others. It is for this reason that the highly anisotropic materials have recently been described with a theory originally developed to describe materials consisting of superconducting layers separated by non-superconducting material. The Lawrence-Doniach theory[69] states that the superconducting properties of a material may be described as existing within the layers, which are coupled weakly to the other layers via Josephson tunneling. (If the variation of the superconducting properties along c is slow compared to the c-axis coherence length ξ_c, then an effective mass tensor may be employed in the description. However, the effective mass idea is not essential to the theory.) The vortices are usually pictured as "pancakes" within the superconducting plane and weakly coupled to the others in neighboring planes. The description has now gone from the 3-dimensional idea of tubes to a 2-dimensional idea of pancakes. This theory has been used recently with fair success in describing some of the more anisotropic materials[53, 70, 71, 72]. It is believed that the anisotropic London theory breaks down in these materials at low temperatures due to the c-axis penetration depth becoming shorter than the spacing between CuO_2 planes.

4.9 Favorable Properties of YBCO

It would appear, with all of the possible effects of anisotropy coupled with the limitations of the London theory, that the theory as developed has no application. This is not the case. The compound YBCO seems perfectly suited to a description by the uniaxial, anisotropic London theory. The four properties which makes it suitable are now briefly described.

The first property YBCO displays is a high-κ value. It is generally believed that the κ value for YBCO is about 100[73]. This certainly places YBCO in the high-κ limit discussed above and therefore makes possible the application of the London theory.

YBCO is anisotropic, so the anisotropic London theory is applicable. However, it is believed to have a $\Gamma = 25$, so the amount of anisotropy is not so great that a layered description is needed to describe the behavior[70, 72]. Hence the London theory as described above should still be applicable.

In YBCO there exists a low-field regime where there is a well developed FLL without too much distortion[55, 74]. There is the possibility of forming a chain state[68], but for fields which are not too low and angles θ which are not too large the vortex state should dominate[58].

Lastly, there has lately been a plethora of good quality single crystals made of YBCO.

These have displayed low twinning and pinning[58, 75]. Crystals of this caliber are essential if a description based on London is to be used.

These four properties of YBCO should allow a low-field study in good single crystals where a well defined, trianglar FLL exists. The London theory should be directly applicable to a YBCO single crystal in this state, and the development in the remainder of this work assumes such a condition can be obtained.

Chapter 5

Magnetic Field Surfaces and Distributions

This chapter first describes the numerical results of the application of the anisotropic London theory to a material like the superconductor YBCO. Three dimensional plots of field structures for the various components of the magnetic fields are shown and described. Contour plots of these surfaces are also shown. Second, the method of obtaining a distribution from any three dimensional surface is described, and the particular application to the magnetic surfaces is shown. Lastly, a discussion of the various aspects of the field distributions is given.

5.1 Magnetic Field Surfaces

The prescription described in the last chapter has been followed and magnetic fields at points within an equilibrium FLL have been calculated numerically. The program needs values for the following parameters: B - the average magnetic field; λ - the effective penetration depth; θ - the angle between the average field and the crystalline **c** direction; and Γ - the anisotropy parameter from which the effective masses m_1 and m_3 are found.

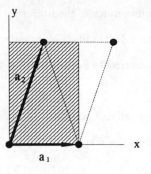

Figure 5.1: The area within the FLL used in calculating the surfaces shown below.

Using only these four parameters the program calculates the real and reciprocal lattice basis vectors, calculates the effective mass tensor, sums over the appropriate number of reciprocal lattice vectors, and transforms back into real space to give the magnetic field components at the point.

The points at which the field is calculated are determined within the program, and lie in a region of the FLL as depicted in Fig. 5.1. This area was chosen as the minimum in which the *vector* nature of the fields is not repeatable. That is, even though the magnitudes of the fields may be reproduced within this area, the vector nature is not. This can be more easily understood by looking at Fig. 4 of reference [46], where the streamlines of the transverse field are shown. The reason for this choice of area will become clear later when the ideas of the off-axis fields are introduced. The actual points within the area where the fields are calculated are on a grid 51x51, for a total of 2601 points. The grid lengths are found by first calculating the real space basis vectors of equation 4.21 and dividing $\mathbf{a_1}$ and the y component of $\mathbf{a_2}$ by 50.

The values of the parameters for $YBa_2Cu_3O_7$ are: $\Gamma = 25, \lambda = 256.5 \ nm$ [44]. The value for λ is found from using $\Gamma^{1/2} = \lambda_c/\lambda_{ab}$, $\lambda = \left(\lambda_c\lambda_{ab}^2\right)^{1/3}$, and a value for $\lambda_{ab} = 1.5 \times$

10^{-5} cm, which is in the range found in much of the literature [17, 76, 55, 77, 78, 39]. The symbols λ_c and λ_{ab} are the penetration depths for fields oriented parallel and perpendicular to the basal planes, respectively (or,currents flowing parallel to the c and a-b directions, respectively).

Results of calculations using a similar λ parameters will be shown. The reason for using a different λ parameter is that the results for the distributions (calculated later) show a more interesting structure than for the more "exact" YBCO parameter. The value used here is $\lambda = 120$ nm, which is about one half of the value quoted above for YBCO. Using this value, $B = 250$ G, $\theta = 45^{\circ}$, and $\Gamma = 25$, the following set of figures was created. Figures 5.2 and 5.3 are the surfaces corresponding to the z component and the magnitude of the field b within the FLL area. Figures 5.4 and 5.5 are the surfaces corresponding to the x and y components of the fields within this area of the FLL.

Figs. 5.2 and 5.3 are almost identical, and have obvious tall, column-like features which are the flux tubes. These tubes extend higher than shown in the figures, but they have been graphically cut off in order to show the field behavior near the central region with more clarity. The fields fall off smoothly from the cores and become relatively flat in the mid-region. The minimum field exists near the center of the plot. In addition, there are three saddle points within the surfaces. Two are equivalent and exist on the diagonals between the cores, and one exists along the x direction between the cores. These saddle points will become very important later when the field distributions are discussed.

Fig. 5.4 is the surface of the x component of the fields. Its shape is almost opposite to the shapes just described for the b_z and b surfaces. Now the surface descends steeply at the cores, becoming negative. The central region again is mainly flat, and the values of the field x component are positive here.

Fig. 5.5 is the y component surface of the FLL. This surface exhibits interesting behavior near the cores, where the fields have an inflection as one passes through the core

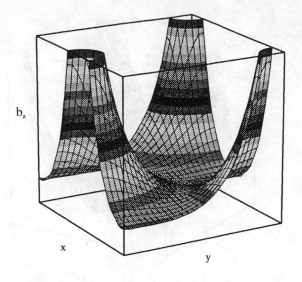

Figure 5.2: The surface corresponding to the z component of the fields.

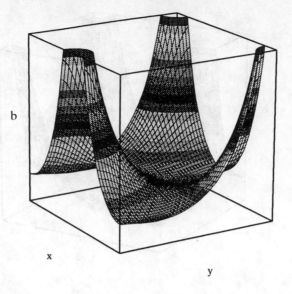

Figure 5.3: The surface corresponding to the magnitude field b at points within a unit cell of the FLL.

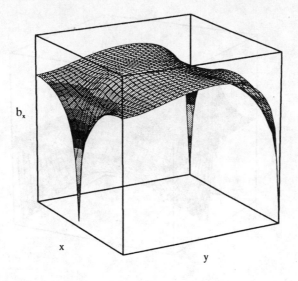

Figure 5.4: The surface corresponding to the x component of the fields.

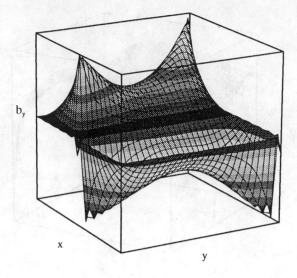

Figure 5.5: The surface corresponding to the y component of the fields.

in the x direction. There is an obvious asymmetry with respect to a plane half way along the x axis, where one side is exactly inverted with respect to the other.

An interesting point concerning both of these last two surfaces is that the average of each over the FLL unit area must be zero. This is almost immediately obvious when one looks at the b_y surface, but is somewhat less apparent when looking at the b_x surface. The reason is rather simple. The average field lies along the z direction, and therefore the dominant field is the longitudinal, or z component. For the average field to be along z, each of the x and y fields, when averaged over the FLL area, must be zero, or else the average field would lie in some other direction. This assertion has been used as a check on the reliability of the surfaces produced with the program. Both the x and y surfaces have been shown to average to zero over the FLL area to within numerical accuracy.

Shown next are the contour plots corresponding to these surfaces. Fig. 5.6 is the contour for the b_z surface. Once again the core regions are graphically cut off so that the region away from the cores may be viewed more clearly. The minimum is clearly visible near the center of the plot. The saddle points are more evident on this plot; one can clearly see by the numbers labeling the contours that there is a saddle point on the line midway between any two cores. Note also that this plot, as well as the two which follow, is drawn with the proper aspect ratio so that the isosceles nature of the vortex lattice can be seen.

The next figure is the contour plot of the b_x surface – Fig. 5.7. The great variation in the values of b_x near the cores is evident in the tight packing of the contour lines there. The flatness away from the cores is also evident, but here some subtle field variation – islands – which was not so obvious on the surface plot, is more easily seen.

The next contour plot is Fig. 5.8, the contour plot of the b_y surface. The symmetry mentioned above is apparent at first glance here, also. From the numbers one can easily see that the two sides are simply inverted reflections of each other. It is also very easily

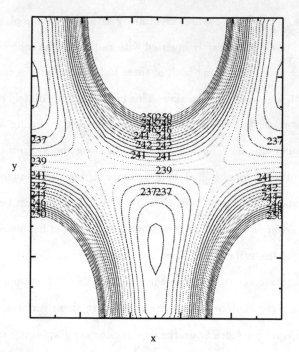

Figure 5.6: The contour plot corresponding to the z component surface. Notice the minimum near the center as well as the saddle points. The core regions have been excluded because the contours became unresolvable. The numbers on this and the following plots are field contour values in Gauss.

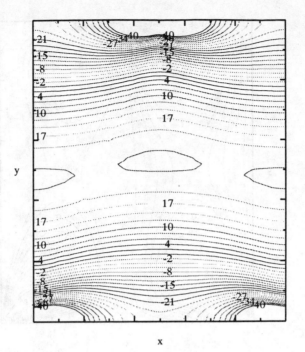

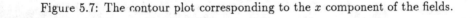

Figure 5.7: The contour plot corresponding to the x component of the fields.

understood from this plot how the fields average out to zero over the area.

It is also possible to calculate the surface of the transverse field $b_\perp$ and show its contour. This has been done by the author, but is not particularly illuminating. More informative is the plot of the lines of the transverse field, often called the *streamlines of the transverse field* because the direction of the transverse field at points is given and it appears as though the lines are flowing. The calculations for these plots has been done by various authors, and two particularly good examples showing interesting streamline plots are references [79, 46]

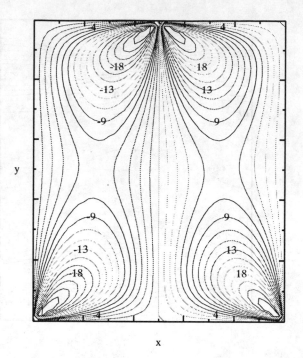

Figure 5.8: The contour plot corresponding to the y component of the fields.

5.2 Method of Calculating Distributions from Surface Grids

This section describes the method of integration employed to allow the distribution of magnetic fields to be calculated. (As a reminder, a distribution is simply a probability of a particular item being found within a certain set of data.) This section is rather tedious and is included for completeness. Uninterested readers may skip this section with no loss of understanding or context.

Calculations of the sort discussed above often leave one with a lot of data and therefore a real need to organize and display it in a meaningful way. Plotting is of course the first option, both as a surface and as contours. This can shed much light on the correctness of the calculations, assuming that the user has some idea of the final result. It can also help in giving some insight as to what the behavior of the particular physical system at hand is doing – the averaging to zero of the b_x and b_y surfaces, for example. It is for these reasons that plotting is of particular importance.

In addition, it is often interesting to determine the distribution of values contained within the grid; or, putting it more mathematically, to calculate the probability distribution of values of the quantity contained within the matrix. Our grids contain magnetic field values which (theoretically) are the actual values of the fields within the FLL. It was discussed in Chapter 2 how the muonic precession signal obtained in a μSR experiment can be Fourier Transformed to yield the distribution of fields that the muons sense. It would therefore be useful to devise a technique to obtain the theoretical field distributions from the grids for comparison to experiment.

The technique described below does just that. It is not unique, and has been used by others to obtain similar kinds of results for things like densities of states[80]. The method does a numerical integration of the following integral:

$$n(M) = \int_A \delta\left(M(x,y) - M_o\right) \, dA \tag{5.1}$$

where A connotes the area of the region of interest (in our case the FLL area), $M(x,y)$ is the value of the quantity contained within the matrix at position (x,y), M_o is the particular value of the quantity at some point of the integration, and δ is the Dirac delta function. For our surfaces the M values will be things like b, b_x, etc. Physically, what this equation does is integrate over contours, finding the fractional amount of area the lines take up within the FLL area, and weight that with the inverse of the steepness of the area on which this particular contour is found. This means that longer contours will contribute more, but if they exist on very steep terrain they are weighted back down.

This can be seen more easily if one uses the properties of delta functions to express the integral in a more physically meaningful form. For a one dimensional function the following relationship is true[81]:

$$\delta(f(x)) = \sum_i \frac{1}{\left|\frac{df}{dx}(x_i)\right|} \delta(x - x_i) \tag{5.2}$$

where the function $f(x)$ has simple zeroes at x_i in the sum. The extension to a two dimensional function is obvious:

$$\delta(f(x,y)) = \sum_i \frac{1}{|\nabla f(x_i, y_i)|} \delta(x - x_i)\delta(y - y_i) \tag{5.3}$$

Replacing the delta function over M in equation 5.1 with the form shown in equation 5.3 yields:

$$n(M) = \int_A \sum_i \frac{1}{|\nabla M(x_i, y_i)|} \delta(x - x_i)\delta(y - y_i) \, dx\,dy \tag{5.4}$$

where of course $dA = dx\,dy$. Evaluation of the integral now leaves a sum over the i places in the area where x and y meet the delta condition:

$$n(M) = \sum_i \frac{1}{|\nabla M(x_i, y_i)|} \tag{5.5}$$

67

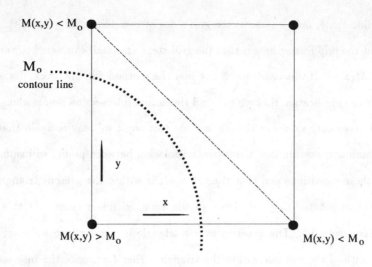

Figure 5.9: A sketch of how the grid is handled for the numerical integration. The value of interest is M_o, and its associated contour line is shown as the darker dotted line. The lighter, diagonal dotted line is that which separates the grid into the two triangles. A given contour can only intersect two of the three sides of each of the two triangles.

The technique works as follows, with reference to Fig. 5.9. The grid is made up of uniformly spaced points in two dimensions. The step sizes in each direction are in general different, but that is irrelevant for the present discussion. The only thing necessary is an NxN grid of values which are relatively smoothly varying in both directions. This can be checked by plotting and is certainly true in the case of our surfaces.

One first chooses a value for M_o and scans through the grid until a specific criterion is met – see Fig. 5.9. The criterion for this check is that the value M_o lie between the minimum and maximum values of the three grid points making up a triangle which is formed by slicing the grid cell along a diagonal – i.e. $M_{min}(x,y) < M_o < M_{max}(x',y')$. The reason for the triangle is that it contains the minimum amount of grid points needed to completely determine the field within its area. If the condition is satisfied, one next finds the two points where the M_o value actually intersects the sides of the triangle. Beware: there is some implied knowledge here. First, it is assumed that the "contour" corresponding to M_o only intersects the grid sides at two points. In general this may not be true, but the implication here is that the grid steps are small compared to variations in the values $M(x, y)$. If this condition is not met the method will not work. Hand-in-hand with this is the implication that we can find the actual intersection points along the sides of the grid given only values at the corners. At this point we require again that the step sizes are small and assume that a linear extrapolation between points will suffice.

Given these conditions and that the value M_o is within the present triangle, we find the intersection points using the slopes of the assumed linear sides. At this point we can apply equation 5.5. The gradient over a triangle is constant, so we must now sum over the length of the contour within the triangle. Therefore, once the intersections are found, we calculate the length of the line connecting the two points – l. Here again is the assumption that the "contour" is slow enough in its variation to allow a linear approximation. In addition to the length we also find the gradient as:

$$\nabla M(x, y) = \frac{\partial M(x, y)}{\partial x} + \frac{\partial M(x, y)}{\partial y} \rightarrow \frac{\Delta M(x, y)}{\Delta x} + \frac{\Delta M(x, y)}{\Delta y} \tag{5.6}$$

where the $\rightarrow$ indicates the partials going over in the linear approximation to differences. The actual contribution to the total probability for a given triangle is:

$$n_\Delta(M) = \frac{l}{\frac{\Delta M(x,y)}{\Delta x} + \frac{\Delta M(x,y)}{\Delta y}} \tag{5.7}$$

The contribution to the distribution for this value of M_o for this particular triangle is the ratio $l/\nabla M(x, y)$. This gets added to all other contributions from all other triangles for the given M_o value. This is then the probability of finding a given M_o within the surface. The next value of M_o is then used and the process is repeated, building up the entire distribution.

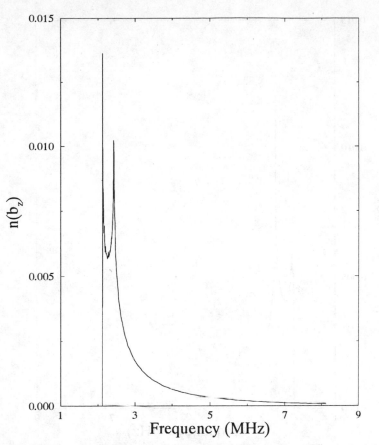

Figure 5.10: The magnetic field distribution for the b_z surface. The parameters used to generate this and the following distributions are the same as those used to generate the surfaces and contours: $B = 250$ G; $\lambda = 120$ nm; $\theta = 45°$; $\Gamma = 25$.

5.3 Magnetic Field Distributions

The method just described above has been implemented and applied to various magnetic field surfaces. The results of this implementation to surfaces like those shown previously in Figs. 5.2 - 5.5 are shown below in Figs. 5.10 - 5.13, respectively.

Figs. 5.10 and 5.11 represent the field distributions for the b_z and b surfaces, respectively. These distributions look very similar, as do their field surfaces in Figs. 5.2 and 5.3. These distributions are characterized by the number of distinctive steps and discontinuities in the curves. For example, both have zero probability for fields below a certain

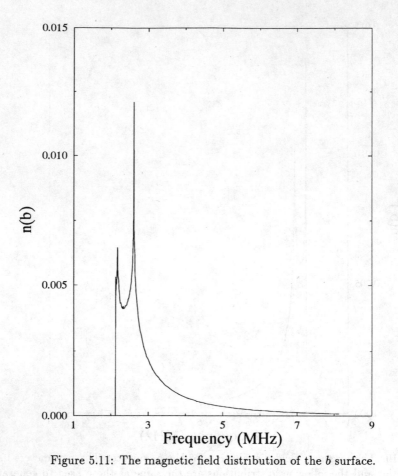

Figure 5.11: The magnetic field distribution of the b surface.

minimum value, which suddenly appears with a very finite probability. This feature is due to the minimum near the center of the surface, which lies in a region which is very flat and therefore has a considerable contribution from the $1/\nabla M(x, y)$ part of the sum (equation 5.7).

Another feature which is characteristic of these surfaces is the peak just above the minimum – which in this case is two peaks. The peaks here are due to the saddle points contained within the field surfaces. These were noted above when both the surfaces and contour plots were discussed. The saddle point along the x axis of Fig. 5.6 is the cause of the left peak, while the saddle points along the diagonals connecting the cores are equivalent and cause the right peak. This can be seen by checking the field values which are printed on the contours. Once again the flatness of the surface around these regions is the cause of the peakiness.

In general there is one further feature common in these particular distributions, and that is a step similar to the minimum at the maximum field value. This step is greatly reduced in magnitude from the minimum, but is likewise caused by the similar effect of there being no field values after it. The reduced magnitude is due to the higher fields existing near the core regions on the surfaces and thereby having shorter lengths and larger gradients contributing to equation 5.7. On the plots in Figs. 5.10 and 5.11 this upper step is not visible. This is because the vertical scale of the plots is large compared to the amplitude of the upper cutoff. (Technically speaking, since the London theory has no intrinsic cutoff, the high field tail will continue on forever. Numerically, however, the reciprocal sum *is* cutoff and the core value is finite, producing a step at some value b_{max}.)

The field distribution for the b_x surface is shown in Fig. 5.12. It has a highly asymmetric shape and also an asymmetry with respect to the origin. In contrast to the distributions just described, this has a sharp, very finite maximum field discontinuity. This is due to the very flat region in Fig. 5.4, where the maximum field situation is almost exactly the

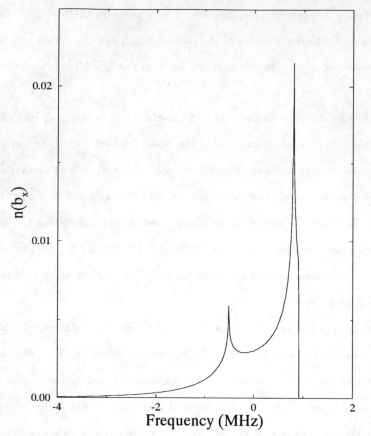

Figure 5.12: The magnetic field distribution corresponding to the b_x surface. Note that the area to the left of the origin is equal to the area to the right – confirming the areal average of zero.

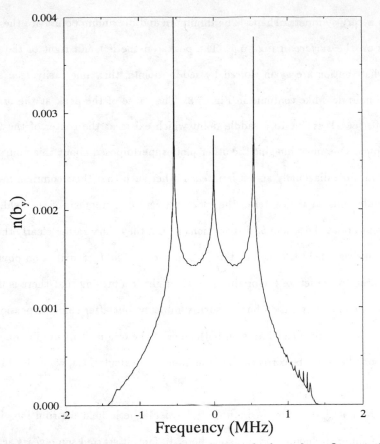

Figure 5.13: The field distribution corresponding to the b_y surface. Once again the areas to the left and right of the origin are equal.

same as for the minimum above. The minimum here behaves as the maximum above, with a long tail seeming to be asymptotic to the x axis at low fields. Here, as with the maximum above, there exists a very small discontinuity or step corresponding to the absolute minimum in the distribution. It is not visible here for the same reason that the maximum step is not visible above. The third feature which catches the eye for this plot is the sharp peak just below zero frequency. This is due to the saddle point most easily seen in Fig. 5.7 at the center of the x axis, and also existing equivalently in both the upper left and right corners of the plot.

The b_y field distribution is shown in Fig. 5.13. In contrast to the other distributions shown it has a very symmetric shape. The minimum and maximum points are the result of the peaks on the *surface* in Fig. 5.5. The peaks on the left and right of the origin in the field distribution are again caused by saddle points, this time easily seen as the center of the hourglass-like contours in Fig. 5.8. The cause of the peak at the origin is not so easy to see. It is due to a saddle point which exists at the center of the top of the contour plot. One must imagine the other plots superimposed about this point, with maxima and minima diagonally across from each other such that their common meeting point is a saddle point at the center of the core with zero y component of the field.

A final point about the b_x and b_y distributions is that they show rather clearly the idea of zero net transverse field throughout the FLL unit cell. The b_y distribution obviously has as much area to the left of the origin as on the right, confirming that there is no net y component. The b_x distribution has no such symmetry, but after some inspection one can see that in fact there is equal area on both sides of the origin. This prediction of the theory is a good check on the correctness of the numerical calculations, which in this case seem to work well.

It should be noted that the theoretical study of magnetic field distributions in the mixed state of superconductors did not start here. In fact, it started some years ago and has evolved quite rapidly since the introduction of the high speed computer. Listed here for the interested reader are some references which show the development of numerical calculations of magnetic field distributions as well as some which show very detailed theoretical analyses of things like peaks and steps. The references are: [50, 82, 83, 84, 85]. The list is not complete, but is a good launching point for anyone wishing to know more.

5.4   Field Distribution Dependence on Parameters

The surfaces, contours, and the corresponding distributions shown are representative of the infinite number which can be obtained by varying the various parameters of the theory.

These particular surfaces were chosen as candidates because the parameters used show a rich variety of structures with interesting behavior. As mentioned before, the only real difference is in the parameter λ, which was used above as 120 nm and should really be 256.5 nm for YBCO. (The real YBCO values will be used below in the discussion of the off axis fields.) Since it has already been hinted that varying a parameter affects the results, we will now discuss just how the distributions and surfaces in fact do vary with the parameters.

5.4.1 Angle Dependence

The angular dependence of the field distributions is shown in Figs. 5.14 through 5.15. In all of the figures in this subsection, the parameter values other than θ are: $B = 250$ G, $\lambda = 256.5$ nm, $\Gamma = 25$. First, the dependence of the longitudinal, or z, component of the fields is shown in Fig. 5.14. The distribution becomes more compact as θ increases to 90°. The minimum field value increases with theta until the left side of the curve becomes a straight, vertical rise to the peak. This means that the minimum and the saddle points have become indistinguishable. The peak also moves toward higher frequencies, and the long high-frequency tail becomes shortened. This behavior indicates that the overall range of b_z within the FLL unit cell decreases rather markedly with angle.

In addition to this behavior is the appearance of the second peak for $0^o < \theta < 90^o$. At $\theta = 0°$ the FLL is isotropic, and hence the saddle points are equivalent in field value. As theta increases, the fields slowly distort (due to the anisotropic masses) and the FLL changes over from equilateral to isosceles, causing differing values for the two saddle points. The FLL remains isosceles all the way to 90°, but here the second peak disappears. In theory the two saddle points are still distinct, but their values have become so close at this angle that one needs to look extremely closely to resolve them. It is worth noting that multiple peaks of any nature have yet to be found experimentally, let alone for the difficult 90° case.

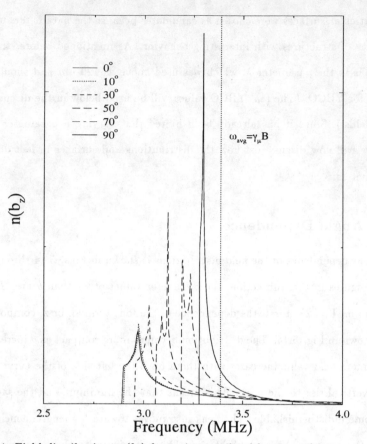

Figure 5.14: Field distributions $n(b_z)$ for various angles θ between the average field $B = 250\ G$ and the crystal Z direction. The other parameters are: $\lambda = 256.5\ nm$, and $\Gamma = 25$. The vertical line at $f = 3.3875\ MHz$ indicates the position of the average field.

A final point on this graph is to note that the variations are all about the average field value, which for a field of 250 G is a frequency of $f = 3.3875 \ MHz$ for positive muons (recall $\omega = \gamma_\mu B$). It is also important to note that the peaks are below the average field value for all angles, as can be seen in the figure where the average field is labeled as a vertical dotted line.

Very similar behavior is shown in Fig. 5.15, where field distributions of the magnitude b are plotted for various angles θ. The same minimum and tail behavior is seen as in Fig. 5.14, as well as the overall decrease in the width of the distribution. The actual values of the minima and peaks are of course slightly different from those in the previous figure, but the angular dependence is the same. However, the distributions at $0°$ and $90°$ are the same as in previous figure, once again because of the lack of transverse components at angles where the average field is aligned with a crystal axis. In this figure the frequency corresponding to the average field B is again labeled as a vertical dotted line. The behavior of the various distributions with respect to this line mirrors that in the last figure.

Somewhat similar behavior is shown in Fig. 5.16, where the b_x distributions are shown for various angles θ. The distribution shrinks dramatically in width as theta is increased, and (due to areal conservation) becomes much more pronounced vertically. The saddle point peak on the left is almost gone at low angles, but moves left and increases as θ gets large. Notice in the graph that the angles $0°$ and $90°$ are not shown; this is because there are no transverse components at angles where the field B is along a crystal direction.

The distributions of b_y are shown in Fig. 5.17. The same angular behavior is observed here as before, with the distribution becomming much more compressed as θ increases. The gradual, symmmetric, wing-like decreases to zero at lower angles become sheer drops at higher angles. The three peaks, which are very widely spaced at $10°$, become almost indistinguishable at an angle of 70 degrees.

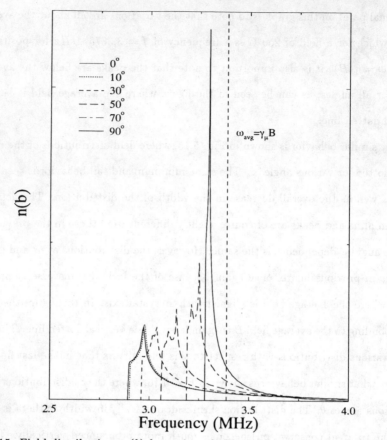

Figure 5.15: Field distributions $n(b)$ for various θ. The other parameters are the same as in the last figure.

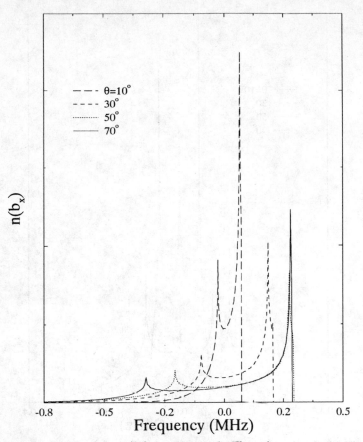

Figure 5.16: Field distributions $n(b_x)$ for various θ. The other parameters are the same as in the last two figures.

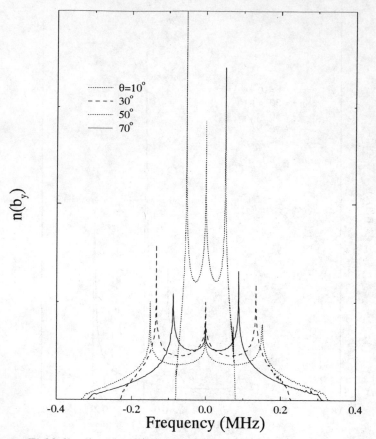

Figure 5.17: Field distributions $n(b_y)$ for various θ. The other parameters are the same as in the last three figures.

In sum, the distributions are all similarly affected by variations in the direction of the average field B. All show significant narrowing as theta increases to higher angles, which is likely a reflection of the increasing "distortion" of the FLL away from equilateral triangles at $\theta = 0°$ to ever more isosceles triangles as $\theta \rightarrow 90°$. (The reader should note the use of the term "distortion" in this context means the natural moving of the FLL from equilateral to isosceles, and *not* distortion by any means such as pinning or thermal fluctuations). The distributions for $n(b_z)$ and $n(b)$ are very similar (even at $B = 250$ G), and show for these parameters separate peaks for the two *different* saddle points within the field surfaces. These peaks arise once theta is non-zero and persist until theta nears 90°, when they disappear again to within the limits of numerical accuracy.

5.4.2 Field Dependence

The dependence of the various distributions on the magnitude of the average field B is discussed here. In all plots the parameters other than B are held fixed at the following values: $\lambda = 256.5$ nm, $\theta = 70°$, and $\Gamma = 25$. The values for B are 100, 250, 500, 1000, and 5000 G.

The $n(b_z)$ plots are in Fig. 5.18 and again show the characteristic shape, but are seen to be displaced with respect to one another. The displacement as shown is not real – the horizontal axis is not the correct axis and has been left unlabeled in an attempt to avoid confusion. The actual positions of the lineshapes on this axis are quite far apart. When they are plotted in these positions they cannot be easily compared. For this reason the distributions above 100 G have been shifted down toward the first curve for ease of comparison.

It is seen at first glance that the magnitude of the average field has some effects on the shapes of the distributions. However, the widths and heights are all very similar, especially when possible numerical effects are considered. It is therefore almost entirely

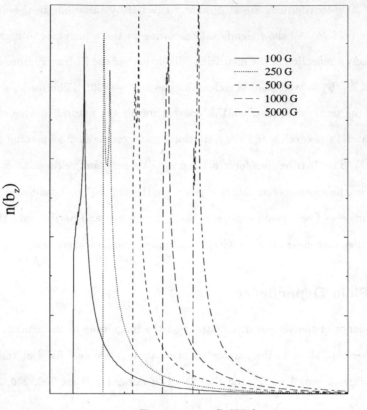

Frequency (MHz)

Figure 5.18: Field distributions $n(b_z)$ for various B. The other parameters are: $\lambda = 256.5$ nm, $\theta = 70°$, and $\Gamma = 25$. Note that the horizontal axis is not real – the curves have been shifted to positions closer together for easier comparison.

correct to state that variations of this sort, especially at intermediate fields, to first order only shift the distribution along the frequency axis.

As mentioned above, there are some differences between the curves. All of the curves, except for the 100 and 5000 G, have some type of double peak structure. These two peaks are caused by the existence of two saddle points within the magnetic field surface (see above). For angles which are intermediate, like $\theta = 70°$ for the figure, one expects the two saddle points to give rise to the peaks. Therefore the lack of a double peak for these two curves is curious, and deserves closer inspection.

For a deeper understanding we go back to the surfaces and inspect the contour plots for three of the curves of Fig. 5.18. These are shown in Fig. 5.19, for the cases where $B = 100, 250$, and 5000 G. The other parameters for these curves are: $\lambda = 256.5$ nm, $\theta = 70°$, and $\Gamma = 25$. Note that the angle here is different than that used earlier for the surface/contour plots; this angle was chosen for two reasons: 1. the geometry of the FLL is different than earlier; 2. this angle is closer to that which yields the maximum saddle point separation – see Fig. 5.14. In Fig. 5.19, it is immediately apparent that the field surfaces undergo changes, especially with respect to the position of the minimum. In the center graph we see a plot like above, where it is easy to see the minima and the saddle points. It is also easy to tell that the two different saddle points (bottom center and along either of the diagonals connecting the lower cores and the upper one) are at distinct field values, separate from each other by a few gauss. When one looks at the left plot one sees that for a lower field of 100 G the minimum has moved down to a position almost colinear with the cores. It's movement has caused the saddle point which resides there at 250 G to be suppressed, because there is now very little area near it which is flat. In the right plot is the contour for $B = 5000$ G. Here again the minimum has moved, but this time it has shifted up toward the top core. If one looks closely one sees that the saddle points are now almost equivalent in field value. In fact, the field surface now has characteristics which resemble quite closely that of a $\theta = 0°$ equilateral FLL surface. There is still a slight difference in the two saddle points, but the difference at 5000 G is already too small to see without integrating over a range which stresses the limits of the numerics.

The behavior of the $n(b)$ curves is almost exactly the same as just described for the $n(b_z)$ curves – see Fig. 5.20. Once again the horizontal axis is not real, and the curves have been shifted to the left as described above. The shape and size of the curves remains almost the same for all B. But, one can also see differences between the curves. Here, like above, there are some curves with multiple peaks. However, unlike above, the multiple

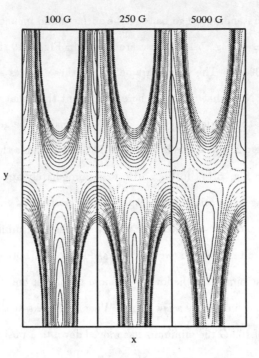

Figure 5.19: Shown above are contour plots of b_z surfaces corresponding to three different average field values. From left to right are: $B = 100,\ 200,\ 5000\ G$. The other parameters are: $\lambda = 256.5\ nm$, $\theta = 70°$, $\Gamma = 25$. Note the movement of the minimum, and its effect on the saddle points, as the average field value increases.

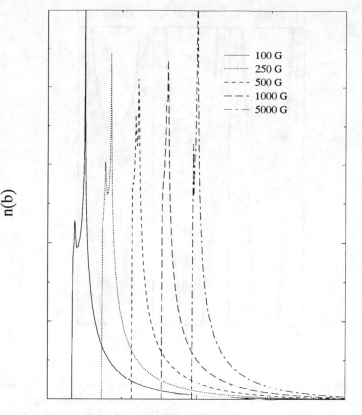

Frequency (MHz)

Figure 5.20: Field distributions $n(b)$ for various B. The other parameters are the same as in the last figure; here, too, the horizontal axis is not real.

peak graphs exist at all fields. Shown in Fig. 5.21 are the contour plots which correspond to those in Fig. 5.19 for b_z. The 100 G plot clearly shows that the minimum is *not* shifted down to the X axis, thereby allowing the saddle point to exist with reasonable area. The 5000 G plot is similar to that in Fig. 5.19, with the distribution looking isotropic. In this case one can see a small peak on the $n(b)$ curve corresponding to this slightly different field-valued saddle point.

We next look at the behavior of the field distributions $n(b_x)$ as the average field magnitude is varied. The curves are shown in Fig. 5.22, where it is seen that the distributions

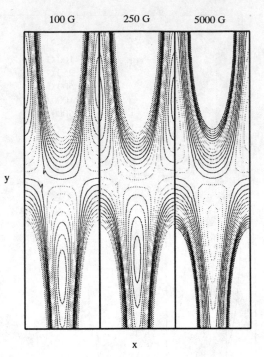

Figure 5.21: Contour plots for the magnitude of b surfaces for $B = 100$, 250, and 5000 G. The behavior of the minimum is similar to the b_z contours, but different for the $B = 100$ G case. This position allows for a more concrete saddle point position and yields a peak on the field distribution curve of the last figure.

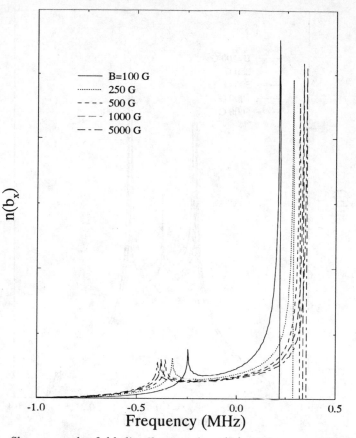

Figure 5.22: Shown are the field distributions for $n(b_x)$ as the average internal field is varied as labeled. The variations here are relatively minor and do not indicate the type of behavior exhibited above for $n(b_z)$ and $n(b)$.

maintain a very consistent shape over a very wide range of average fields. The characteristic asymmetric peaks spread out as the average field is increased. The maximum field value seems to be approaching some upper limit, and the left peak shifts down but also seems to approach a limit. Therefore the values of the fields at the max and left peak saddle do vary with magnitude of the average field. However there appears to be no behavior at all similar to that of the $n(b_z)$ and $n(b)$ distributions, since these saddle points all remain intact throughout the field variation.

Lastly, in Fig. 5.23 are the field distributions for $n(b_y)$. Here there is a very slight

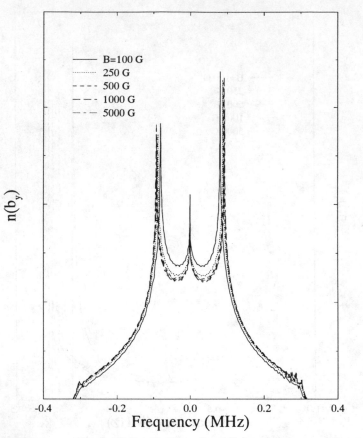

Figure 5.23: Shown are the field distributions for $n(b_y)$ as the average internal field is varied as labeled. The variations here are very minor, and the distributions appear to have reached a limit of change by 1000 G, after which no further variation occurs with increasing field.

variation between the 100 G curve and the others, which are almost right on top of each other. These distributions therefore seem to converge to a limiting form much sooner than the previous ones, and the overall variation with average field magnitude is almost negligible.

In sum, the variation with average magnetic field of the various component distributions is rather slight, especially for the b_x and b_y distributions. The distributions for b_z and b generally retain their shape, but closer inspection reveals interesting behavior of the field surfaces. Specifically, this behavior involves the movement of the minimum on both plots, and on the b_z surface it moves enough to cancel out the effect of one saddle point at $\theta = 0°$ and $\theta = 90°$.

5.4.3 Effective Penetration Depth Dependence

We next look at the dependence of the field distributions on the value of the effective penetration depth, λ. Following the order established above, we will first inspect the distributions of $n(b_z)$. It is interesting first to establish the theoretical dependence of $n(b_z)$ on λ, and then to check that the theory is self consistent. The quantity of interest is $< (b_z - B)^2 >^{1/2}$, which can be derived using equations 4.19 and 4.20. If we assume intermediate field values, we can use the aproximation $\lambda^2 G^2 \gg 1$[46], and the result is:

$$< (b_z - B)^2 >^{1/2} = \frac{m_{zz}}{\lambda^2 m_1} \sum_{G \neq 0} \frac{1}{m_{zz} G_x^2 + m_3 G_y^2} \tag{5.8}$$

The important relation of course is that the width of the $n(b_z)$ distribution varies as $1/\lambda^2$, which we will check below.

The plot of the λ dependence for $n(b_z)$ is in Fig. 5.24, where one can immediately see a very strong variation. These curves have λ values of 100, 256.5, 500, and 1000 nm. The other parameters, in this as well as the other graphs in this section, are: $B = 250\ G$, $\theta = 70°$, and $\Gamma = 25$. The values appear at least qualitatively to have the behavior predicted by the theory.

As a check on the theory, the full width at half max (FWHM) is plotted as a function of λ in Fig. 5.25. Shown as points are the widths, and the curve is a fit of $y = a_0 * \lambda^{a_2}$ to the points. The value of the fit parameter a_2 is -1.9996, indicating that the calculation of the lineshapes is correct (at least as far as the theory holds).

The phrase full width at half max needs a little explanation. Theoretically, the peaks on the curves arise from a singularity and are therefore infinite. Numerically they are finite. The method used in this and the following section for determining the FWHM was as follows. First, the peak was found. Second, an average of many points on either side

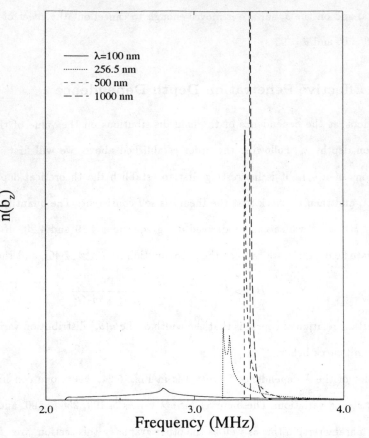

Figure 5.24: Shown are the field distributions for $n(b_z)$ as the effective penetration depth is varied as labeled. The values of the other parameters are: $B = 250\ G$, $\theta = 70^\circ$, and $\Gamma = 25$. The variation in width of the distributions theoretically varies as $1/\lambda^2$.

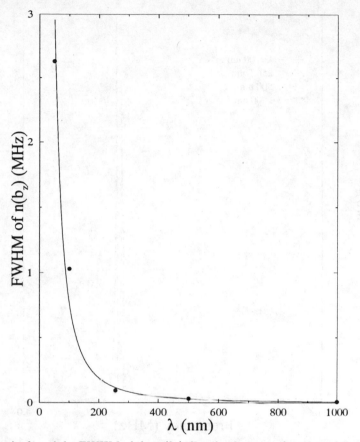

Figure 5.25: A plot of the FWHM of the $n(b_z)$ distributions as a function of the parameter λ. The points are the FWHM of the calculated distributions, and the curve is a fit to these points. Anisotropic London theory predicts a $1/\lambda^2$ dependence, the fit produced an exponent of -1.9996.

of the peak (the higher of the two, if two) was found, being careful not to include points too far from the peak. Third, this average value was then halved, and the width at this level was found. This method worked well for the $n(b_z)$ and $n(b)$ curves, as will be shown below.

The theoretical dependence of the distributions $n(b)$ on the effective penetration depth is the same as for $n(b_z)$. Using the same approximations as above, it can be shown that the FWHM of these distributions also vary as $1/\lambda^2$. The plot of the field distributions for various λ is in Fig. 5.26, and the plot of FWHM for these curves versus λ is in Fig. 5.27.

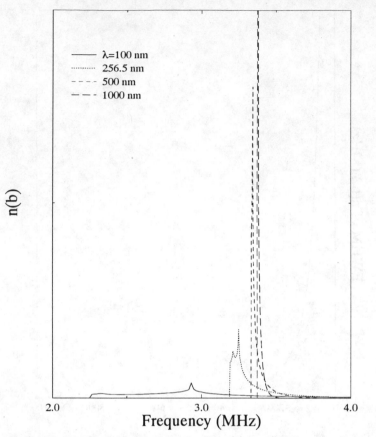

Figure 5.26: Shown are the field distributions for $n(b)$ as the effective penetration depth is varied as labeled. The values of the other parameters are: $B = 250\ G$, $\theta = 70°$, and $\Gamma = 25$. The variation in width of the distributions theoretically varies as $1/\lambda^2$.

The solid curve in the latter figure is another fit to the data, with the exponent this time having a value of -1.995.

The λ dependence of the distributions $n(b_x)$ and $n(b_y)$ are shown in Figs. 5.28 and 5.29. Once again a very strong dependence on λ is evident. Here, we look for the width of the distribution as a variation with respect to zero Gauss $- < (b_x - 0)^2 >^{1/2}$ and $< (b_y - 0)^2 >^{1/2}$. Using similar methods to those used above, we can show that these relationships are given by:

$$< (b_x - 0)^2 >^{1/2} = \sum_{G \neq 0} \frac{B m_{xz} G_y^2}{\lambda^2 G^2 m_1 (m_{zz} G_x^2 + m_3 G_y^2)} \tag{5.9}$$

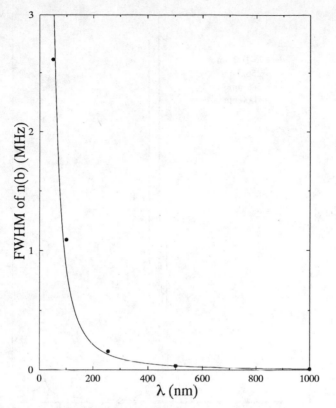

Figure 5.27: A plot of the FWHM of the $n(b)$ distributions as a function of the parameter λ. The points are the FWHM of the calculated distributions, and the curve is a fit to these points. Anisotropic London theory predicts a $1/\lambda^2$ dependence, the fit produced an exponent of -1.995.

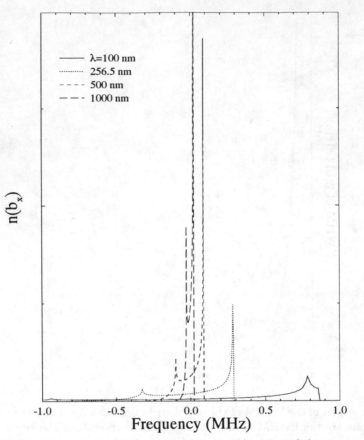

Figure 5.28: A plot of the $n(b_x)$ distributions as a function of the parameter λ.

$$< (b_y - 0)^2 >^{1/2} = \sum_{G \neq 0} \frac{B m_{xz} G_x G_y}{\lambda^2 G^2 m_1 (m_{zz} G_x^2 + m_3 G_y^2)} \tag{5.10}$$

Hence both should also display a $1/\lambda^2$ behavior. A plot of the widths of the $n(b_x)$ lineshapes versus λ is shown in Fig. 5.30. The points show the correct qualitative behavior, but the fit produced an exponent of -1.56 when fit to only the right four points. When fit with all points the exponent was -1.35, which is worse. The widths were determined by the spread between the left peak and the farthest right non-zero point of the distribution. The calculation of the widths was somewhat complicated by the fact that the lineshape form changed at $\lambda = 50 \ nm$. The steep fall on the right side became a more gradual decline, and the two peaks moved closer together. This made defining a consistent width

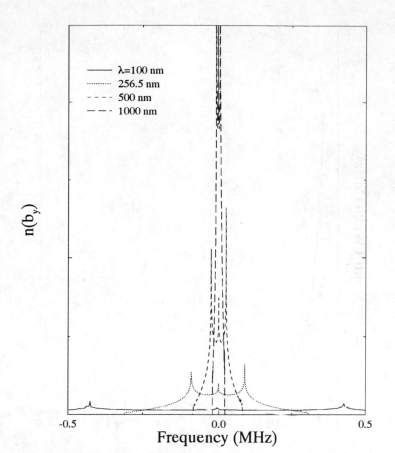

Figure 5.29: A plot of the $n(b_y)$ distributions as a function of the parameter λ.

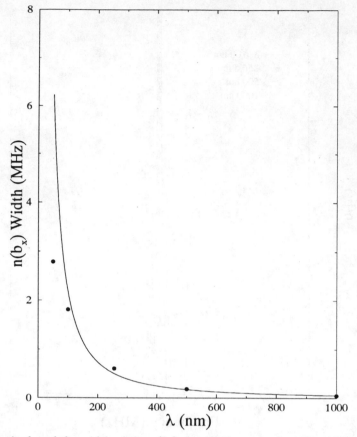

Figure 5.30: A plot of the width of the $n(b_x)$ distributions as a function of the parameter λ. The points are the width of the calculated distributions, and the curve is a fit to these points. The fit produced an exponent of -1.56.

difficult, and hence a fit was done over the final four points to help avoid this problem. The value for the four points is still not great, and this is due to the fact that the 50 nm point is crucial for a more complete mapping of the behavior.

The widths of the $n(b_y)$ distributions were more well-behaved, as shown in Fig. 5.31. The exponent was fit to a value of -1.904, which is fairly close to the expected -2. The widths here were measured between the parts of the curves in Fig. 5.29 where the distribution drops vertically to zero probability. The lineshapes for $n(b_y)$, unlike those of $n(b_x)$, maintain a consistent form for all λ and therefore allow for a more consistent width

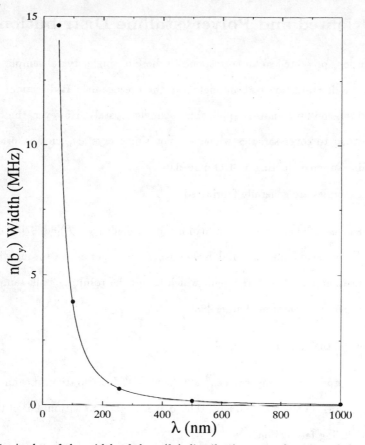

Figure 5.31: A plot of the width of the $n(b_y)$ distributions as a function of the parameter λ. The points are the width of the calculated distributions, and the curve is a fit to these points. The fit produced an exponent of -1.904.

measurement. This is evidenced by the much better behavior of the points and the better fit parameter for the exponent.

To summarize, the anisotropic London theory predicts a $1/\lambda^2$ dependence for the width of the various distributions discussed. Examination of the calculated distributions yields almost exactly this behavior for both the $n(b)$ and $n(b_z)$ curves, very close correspondence by the $n(b_y)$ curves, and a somewhat less than perfect match for the $n(b_x)$ curves. This last is mostly due to difficulties in analysis, and with a more consistent method would probably yield better results.

5.5 Oriented and Polycrystalline Distributions

The distributions presented so far correspond to those of single crystal samples – there is
a known crystalline direction **c** at an angle θ to the average field **B** direction. Therefore
the entire discussion up to now is applicable to single crystals. However, the techniques
can be extended to cover samples which are not single crystals, such as oriented and
polycrystalline superconducting YBCO materials.

Oriented samples are generally fabricated:

1. in a suspension which is placed in high magnetic fields ($\sim 8\ T$)[86, 87]. The crystallites of superconducting material in the suspension experience a torque which tends
   to align their **c** axis with the field, which leaves the resulting bulk sample with a
   high degree of orientational order [88].

2. by melt texturing a thick film[89].

The materials produced in this way generally have $\sim 90\%$ orientation, meaning that the
crystallites making up the material have their **c** axes aligned to within $\pm 5°$ of that of the
applied field during fabrication.

Polycrystalline samples are generally produced from precursor powders which are
mixed in the proper proportions and pressed into pellets. These pellets are then reacted in
a furnace with oxygen, bringing the oxygen level up to that required for superconductivity[90,
91, 92]. In these samples there is no unique crystalline direction – the crystallites are
isotropically oriented in space.

The extension of the field distribution calculations to these types of materials is simple.
One calculates many single crystal distributions $n_{sc}(b, \theta)$ for various angles and then sums
and averages appropriately. This is expressed mathematically as:

$$n(b_z) = \int_{\theta_{min}}^{\theta_{max}} n_{sc}(b_z, \theta) \sin \theta d\theta \qquad (5.11)$$

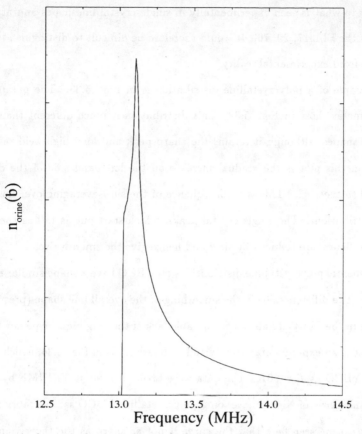

Figure 5.32: A theoretical oriented sample field distribution for a material like YBCO: $B = 1000\ G$, $\lambda = 256.5\ nm$, $\Gamma = 25$. The degree of orientation is roughly 90%, meaning that the crystallite **c** axes fall to within $\pm 5°$ of the intended orientation direction.

where $n(b_z)$ is the oriented or polycrystalline distribution corresponding to the choice of angles θ_{min} and θ_{max}.

For example, one can find the oriented sample field distribution of a 90% oriented sample by setting $\theta_{min} = 0°$ and $\theta_{max} = 5°$. A plot of this calculation is shown in Fig. 5.32. The parameters for the calculation are: $B = 1000\ G$, $\lambda = 256.5\ nm$, and $\Gamma = 25$. This distribution looks similar to the typical single crystal distributions, with the sharp discontinuity at the minimum field, the sharp peak, and then the long, high field tail. The peak is not quite as sharp as before, and the width is slightly enhanced also. These effects are similar to what is seen experimentally in single crystal lineshapes and attributed to disorder in the FLL[17, 29, 76]. It would therefore be difficult to distinguish these subtle differences from experimental reality.

An example of a polycrystalline distribution is in Fig. 5.33. The parameters here are the same as those in Fig. 5.32. This distribution is much different than the single crystal lineshapes, although it retains the sharp peak and long high field tail. The big difference in this plot is the gradual increase on the left hand side of the distribution leading up to the peak. This is a consequence of the $\sin \theta$ averaging over lower θ single crystal distributions. The single crystal peaks, which start out as tall as the remaining peak in the figure, are reduced by $\sin \theta$ and hence give the smooth rise.

Experimental polycrystalline distributions on YBCO have a shape similar to the theoretical one. The differences lie in the smoothing of the overall line shape (presumably due to disorder in the FLL) and also in the notable lack of the long high frequency tail[83, 82]. An example of an experimental polycrystalline lineshape is in Fig. 5.34, which is of a bulk sample of YBCO at $T = 20K$. The data were taken in 1990 at TRIUMF by Dr. Carey Stronach and Christof Niedermayer on a polycrystalline YBCO sample fabricated by the author. It can be seen here that the curve is not as sharp as the theory, and that the high field tail is almost non-existant. In order to make the theory appear more like the data, various schemes have been developed to alter the shape of the theoretical curves. The most popular is convolution with another curve, usually a gaussian, which causes the theoretical curve to broaden and smooth out, like the data [17, 76]. The convolution curve is also shown in Fig. 5.34, but is shown on a reduced scale for clarity.

Other techniques have been developed to alter the appearance of the theoretical polycrystalline lineshapes. All attempt to decrease the theory's high frequency tail and increase the lower frequency side. A few references are included here for the interested reader[93, 94]. These references suggest that the London theory is correct and applicable,

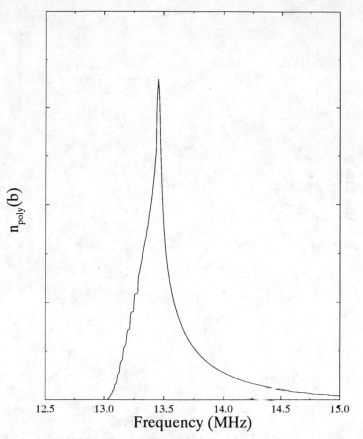

Figure 5.33: A theoretical polycrystalline field distribution for a material like YBCO: $B = 1000\ G$, $\lambda = 256.5\ nm$, $\Gamma = 25$. Characteristic features are the gradual rise on the left and the long high frequency tail.

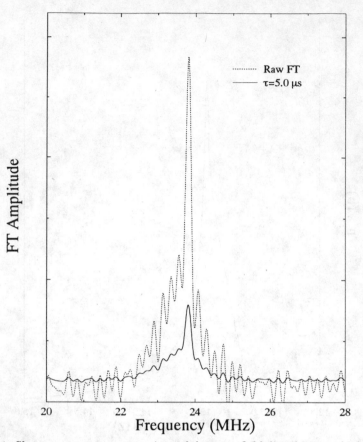

Figure 5.34: Shown are two representations of the same field distribution which are Fourier Transformed from μSR data taken at TRIUMF on a bulk YBCO sample. The temperature of the sample is 20 K. The dotted line is the raw transform, with no tweaking. The solid curve is the result of apodization (multiplication in time space) with a gaussian envelope which has a decay constant of 5 μs – this is essentially convolution.

and that the discrepancies between the theoretical and experimental results are due to things like disorder, bad samples, and pinning.

Chapter 6

Off Axis Fields in Anisotropic Superconductors

In this chapter the microscopic magnetic field components perpendicular to the average field **B** direction are investigated. A numerical simulation of muon behavior within the field distributions of last chapter is presented. The resulting muon time histograms are theoretical predictions of the experimental μSR data. Coupling existing μSR techniques and Fourier transform analysis, the data are treated to reveal what we call *moments* of the microscopic field distributions. These moments tell us interesting things about the nature of the fields within the materials, including an estimate of how far the average field is off in direction from the applied field.

6.1 Simulating Muon Behavior within the FLL

In chapter 4 the anisotropic London theory was developed. It describes a method for calculating the magnetic fields within the FLL on an anisotropic superconductor. In chapter 5 this method was implemented on a computer, allowing the magnetic fields of the FLL to be numerically determined. Using these fields, and our knowledge of how muons

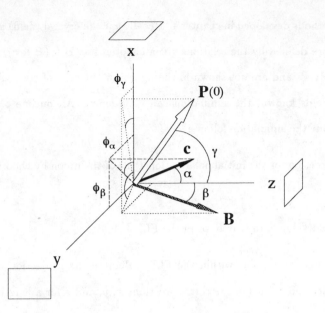

Figure 6.1: The definition of geometry and trigonometry for the discussion in the text. There are detectors along the x, y and z directions indicated by the clear squares.

precess in magnetic fields (chapter 3), allows a simulation program to be implemented where muons stop uniformly within a triangular FLL. The simulation results in data which should correspond to high statistics μSR data. The manipulation of this data will be described below.

One must first understand the geometry used for the simulation. The geometry and trigonometric symbols used are shown in Fig. 6.1. Particular values of the angles will be referenced later when simulation results are discussed.

First, the directions x, y, and z are those which line up with three mutually perpendicular detectors – the LAB axes. All further directions are defined with respect to these three. The average field $\mathbf{B}$ direction is determined by the angles β and ϕ_β. The $\mathbf{c}$ axis direction is determined by the angles α and ϕ_α. Once $\mathbf{B}$ and $\mathbf{c}$ are specified, the angle θ between them is determined by the dot product: $\cos\theta = \mathbf{B} \cdot \mathbf{c}/B$. Knowing θ, the magnitude of $\mathbf{B}$, and the material (*i.e.* λ and Γ) allows the calculation of the fields within the

FLL by the methods developed in chapter 4. Recall that the crystal (field) axes directions X, Y, and Z are defined by the relations given chapter 4 as $\mathbf{B} \times (\mathbf{B} \times \mathbf{c})/B^2$, $\mathbf{B} \times \mathbf{c}/B$, and $\mathbf{B}$, respectively, and are not shown in the figure for the sake of clarity.

With the fields known, the simulation can commence. An outline of the program which carries out the simulation follows:

1. Pick a direction for the initial polarization $\mathbf{P}(0)$ of the muon by choosing the angles γ and ϕ_γ.

2. Pick a field value b_i to search for in the FLL.

3. If b_i is found at a point within the FLL, calculate the projection of the muon's polarization along each of the three directions x, y, and z for a length of time T as it precesses about the vector local $\mathbf{b}$.

4. Using the integration method of chapter 4, weight each muon's contribution according to the amount of area that the field occupied within the cell.

5. Keep a running field/areal average for each of the three mutually perpendicular directions where one has imagined the detectors to be.

The details of point 3 above deserve some discussion. Recall from chapter 4 that the FLL unit cell is divided into rectangular sub-cells. The field values of the FLL are calculated at the corners of the sub-cells. Once a field $\mathbf{b}_i$ is found to lie within a sub-cell, its direction must be determined. We only know the field values at the corners of a sub-cell, so to find the direction of the field $\mathbf{b}_i$ at its intersection point of the sub-cell, we average the components at the two closest known points. These averaged components are then used for the direction of the local field. The muon polarization is then broken into components along this average field direction and perpendicular to it. The perpendicular component is further broken into two components, where the x' direction is defined to lie in the plane defined by $\mathbf{P}(0)$ and $\mathbf{b}_{av}$. The x' and y' components then rotate about the $\mathbf{b}_{av}$ direction in discrete time steps for some amount of time T via the equation:

$$
\begin{pmatrix} x'(t) \\ y'(t) \\ z'(t) \end{pmatrix} = \begin{pmatrix} \cos\omega t & \sin\omega t & 0 \\ -\sin\omega t & \cos\omega t & 0 \\ 0 & 0 & 1 \end{pmatrix} \begin{pmatrix} x'(0) \\ y'(0) \\ z'(0) \end{pmatrix} \tag{6.1}
$$

The components x', y', and z' are then transformed into the LAB axes by rotations about the apropriate axes. The rotation matrix for this is:

$$
R = \begin{pmatrix} \cos\delta\cos\phi_b & -\sin\phi_b & \sin\delta\cos\phi_b \\ \cos\delta\sin\phi_b & \cos\phi_b & \sin\delta\sin\phi_b \\ -\sin\delta & 0 & \cos\delta \end{pmatrix} \tag{6.2}
$$

where δ and ϕ_b are the polar and azimuthal angles determining the direction of $\mathbf{b}_{av}$ with respect to the LAB axes.

6.2 Assumptions of the Simulation Program

This model assumes certain things. For instance, it assumes that all muons arrive with their polarization in the same direction. This is largely true, but it should be remembered that real muon beams may be only 90% polarized. This program, like the previous one, assumes that the FLL consists of a well defined, equilibrium set of straight cores which are not distorted in any way. The possible problems with this assumption were discussed above in section 4.7. Further, this program assumes a uniform stopping distribution of muons throughout the FLL. That is, that the muons are as likely to stop in one spot in the FLL as in any other, and that there will be no feature within the FLL or material which will cause any denser stopping of muons than any other. In order to satisfy this assumption it will be necessary to have very good quality samples which are relatively pin-free and uniform. This assumption is generally quite good with respect to the FLL, because the length scale of the FLL is of the order of λ, while the muon is believed to stop near the oxygen in the plane or chain of the unit cell (Fig. 4.4). The length scale of the FLL is therefore something like 150 nm, while the muons are stopping within the unit cells (with a bond length of approximately 0.1 nm[95, 96]) which are at most of order 1 nm. Therefore many muons should stop within the the FLL and sample it uniformly.

The model also assumes no type of interaction of the muon with anything other than the magnetic field, and that the muons do not diffuse throughout the material. Since, as mentioned above, the muons are believed to stop near an oxygen, then there should be no dipole-dipole interaction since oxygen nuclei are spin zero in their most abundant form. Due to the $1/r^3$ nature of the dipole-dipole interaction, no other nuclei will have an influence even close to that of the FLL fields. In addition, it has been shown [90] that the muons do not diffuse in the high-T_c materials, so motion effects should not play a role in the results.

A further assumption is that the detectors are of infinitessimal size and all of equal efficiency. In reality the detectors cover some solid angle in space and therefore the projection is over this area some distance away from the muon. The efficiency of each detector will in general be different, and there may even be differences within a single detector depending on where the positrons are incident. The actual signal in a finite sized detector will therefore be an average of the type of signal calculated here over the solid angle of the detector and also over the efficiency as a function of position on the detector. This type of averaging will cause the signal to become smeared somewhat, but for higher statistics the problem should be less noticable.

Lastly, it must be understood that the simulation calculates the *projection* of the muon polarization simultaneously along all three directions. Thereore each muon contributes to each detector's signal for the entire time T that the simulation runs. This is certainly much different than a real monte carlo simulation, where actual decays with the decay

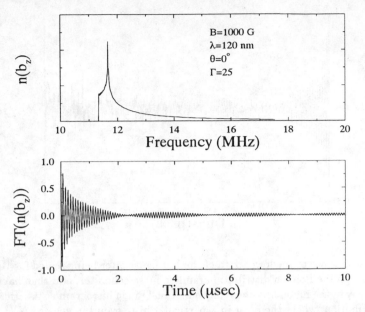

Figure 6.2: Top is a theoretical field distribution for the following parameter values: $B = 1000\ G$, $\lambda = 120\ nm$, $\theta = 0°$, and $\Gamma = 25$. The bottom is the real Fourier transform of this distribution, which corresponds to the time-space or asymmetry representation.

asymmetry folded in would take place at times weighted by the muon lifetime. However, real data will approach our simulation for high statistics.

In order to see the kind of experimental statistics needed to approach our simulations, we did a numerical test using a theoretical lineshape calculated via the methods of last chapter. We first calculated the lineshape from the z surface, and then Fourier transformed into time space to simulate the asymmetry histogram – see Fig. 6.2; top panel is the theoretical distribution, bottom is the FT of the distribution. This data was then transformed into raw histogram-like data by including the muonic decay and a user specified total count parameter. This total count parameter could be varied to simulate the statistics of the histogram – higher total counts meaning better statistics. The now raw histogram was then made noisy by running it through a random number generator [97] which produced a spectrum with a Poisson distribution of noise in it. This noisy spectrum was then changed back to an asymmetry type of spectrum – see Fig. 6.3 – and

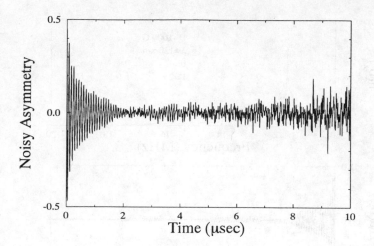

Figure 6.3: This is a plot of the lower data of the last figure after being scaled for 10^6 total events, run through the Poisson distribution random noise generator, and then having the exponential decay of the muon lifetime removed. A real muon histogram's statistics follow a Poisson distribution, where the error in a particular histogram bin goes as $N^{1/2}$, where N is the counts in the bin.

then Fourier transformed into frequency space to see the effects on the lineshape.

The results of three such trials are shown in Fig. 6.4. The amount of statistics is varied by an order of magnitude as one goes from top to bottom from 10^6 to 10^8. The effect of better statistics is obvious, as the original lineshape of Fig. 6.2 is virtually recovered in the bottom plot. We therefore conclude that at *least* 10^7 total events are necessary for an experiment to reasonably reproduce these results.

6.3 Results of the Simulation

6.3.1 Results for a General Case

An example of the results of a simulation are shown in Fig. 6.5. This case has the average field at a position given by $\beta = 0°$, $\phi_\beta = 0°$, the **c** axis at $\alpha = 45°$, $\phi_\alpha = 0°$, and the initial polarization at $\gamma = 80°$, $\phi_\gamma = 0°$. One can see the three separate signals in the figure as

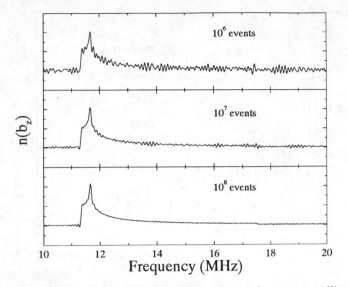

Figure 6.4: Top is the resultant distribution for the total events equalling 10^6. Here the noise is such that the details of the underlying distribution are difficult to determine. The middle panel is for 10^7 total events, where the details of the line shape are becomming more clear. The bottom is for 10^8 events, where most features of the original distribution are reproduced.

different line types. Each shows a realtively large initial amplitude and oscillations which decay at a certain rate. The oscillation frequency closely corresponds to that given by $\omega = \gamma B$, but the signal is actually not simply composed of one frequency, as one might have guessed. The decay of the signal with time is interesting because it is evidence of the existence of the off axis fields. That is, the muons are all arriving in phase due to their polarization, but because each sees a slightly different local field (in both magnitude and direction) each precesses at its own rate in its own direction. Therefore the muons all slowly become out of phase with each other and the net signal seen at the detector decreases. The proof of the existence of the off axis fields is this dephasing in the signal. If there were no off axis fields, then all muons would see the same local field and they would stay together as they precessed. Therefore this dephasing, if all other sources of field inhomogeniety can be ruled out, is definite proof of the existence of off axis fields.

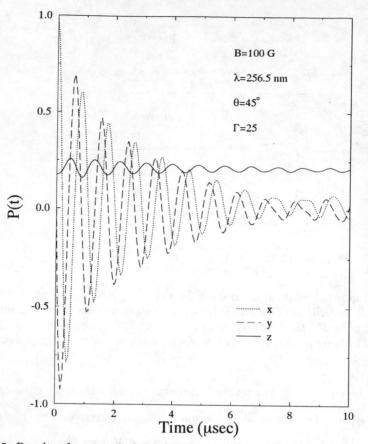

Figure 6.5: Results of a numerical simulation of muons within the FLL of a material like YBCO. The average magnetic field $B = 100$ G and is in a direction given by $\beta = 0°$, $\phi_\beta = 0°$. The crystal **c** axis is at $\alpha = 45°$, $\phi_\alpha = 0°$. The initial polarization is at $\gamma = 80°$, $\phi_\gamma = 0°$. The lines correspond to each detector as labeled in the graph.

6.3.2   Coupling to/from Local Fields

An interesting detail of these simulations is found by observing the behavior of the various signals as the direction of the initial polarization of the muons is varied. For example, given a field value of $B = 100\ G$ in the direction $\beta = 0°$, $\phi_\beta = 0°$, a **c** axis direction of $\alpha = 45°$, $\phi_\alpha = 0°$, and a YBCO-like material, Fig. 6.6 shows what appears to be a phase shift in the signals for two directions of initial polarization. The top panel shows the x signals, the center shows the y, and the bottom shows the z signals. There appears to be a phase shift in both the x and y signals between the $\mathbf{P}(0) \parallel z$ ($\gamma = 0°$) and $\gamma = 20°$. The increase in the initial x amplitude is explainable in that as γ is increased, the projection onto x must therefore increase. But this does not explain the phase change in the y signal.

To investigate further, we calculate the signals as the initial polarization is varied from $\gamma = 0°$ to $4°$. These are shown in Fig. 6.7. We again see what appears to be phase shifts in the x and y signals as the angle γ is varied. When $\mathbf{P}(0)$ is farther from $\mathbf{B}$, the oscillations begin in the proper direction for this geometry via $d\mathbf{P}/dt = \mathbf{P} \times \mathbf{B}$. One can see this by noting that the initial amplitude along the y direction goes negative, corresponding to precession about the z axis in a clockwise sense as viewed along $+z$ back toward the origin. For lower angles of $\mathbf{P}(0)$ there is a definite change in the muon behavior, for now the initial amplitude along y is no longer negative, but positive. It seems that the muon's behavior shifts from one where the precession is governed by the direction of the average field $\mathbf{B}$ at larger angles ($\gamma \geq 3°$) to one which is controlled by the local fields $\mathbf{b}$ near $\gamma = 0°$ – the muons will decouple from the local fields at large γ. It would therefore be interesting to look experimentally for this type of behavior, as it would be yet another indication of the existence of off axis fields.

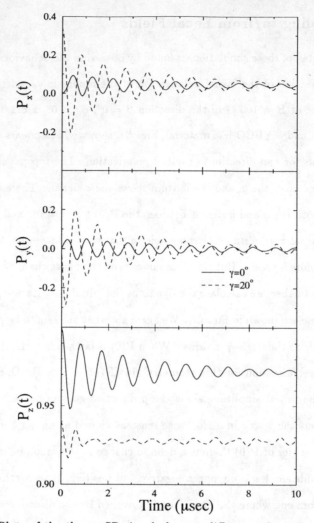

Figure 6.6: Plots of the three μSR signals for two different orientations of $\mathbf{P}(0)$: $\gamma = 0^\circ$ and $\gamma = 20^\circ$. In the top figure there appears to be a phase shift between the two signals. This phase shift also appears in the center plot for the y signals.

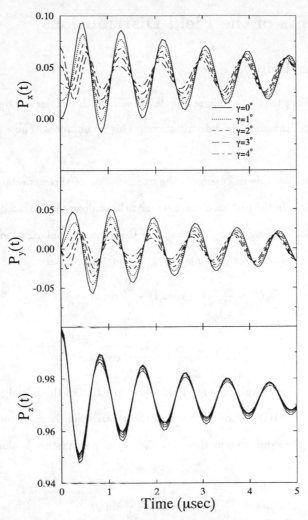

Figure 6.7: The variation in the initial part of the curves as the direction of the muonic polarization is varied through the angles labeled in the figure. Note the phase change in the x and y curves as the angle γ increases from $0^{\circ} \to 4^{\circ}$.

6.4   Moments of the Field Distributions

6.4.1   Starting Point

It was mentioned above that a technique was developed which enables the determination of various moments of magnetic field distributions. This technique will now be discussed in some detail.

The starting point for this discussion is the examination of three special cases of experimental geometries. In the first case, the average field is along the LAB z direction and the initial polarization is along the x direction. For this case, the polarization projections along each of the three LAB axes (corresponding to equation 3.2) are:

$$P_{xy}(t) = \frac{b_y b_x}{b^2}(1 - \cos\omega t) - \frac{b_z}{b}\sin\omega t \tag{6.3}$$

$$P_{xz}(t) = \frac{b_z b_x}{b^2}(1 - \cos\omega t) + \frac{b_y}{b}\sin\omega t \tag{6.4}$$

$$P_{xx}(t) = \frac{b_x^2}{b^2} + \frac{b_y^2 + b_z^2}{b^2}\cos\omega t \tag{6.5}$$

where the first of the double subscripts indicates the direction of the initial polarization, and the second is the LAB direction upon which the polarization is projected.

The next case is the same, except that now the initial polarization is along the LAB y direction:

$$P_{yz}(t) = \frac{b_z b_y}{b^2}(1 - \cos\omega t) - \frac{b_x}{b}\sin\omega t \tag{6.6}$$

$$P_{yx}(t) = \frac{b_x b_y}{b^2}(1 - \cos\omega t) + \frac{b_z}{b}\sin\omega t \tag{6.7}$$

$$P_{yy}(t) = \frac{b_y^2}{b^2} + \frac{b_z^2 + b_x^2}{b^2}\cos\omega t \tag{6.8}$$

The final case has the initial polarization along z:

$$P_{zx}(t) = \frac{b_x b_z}{b^2}(1 - \cos\omega t) - \frac{b_y}{b}\sin\omega t \tag{6.9}$$

$$P_{zy}(t) = \frac{b_y b_z}{b^2}(1 - \cos\omega t) + \frac{b_x}{b}\sin\omega t \tag{6.10}$$

$$P_{zz}(t) = \frac{b_z^2}{b^2} + \frac{b_x^2 + b_y^2}{b^2}\cos\omega t \tag{6.11}$$

The simulations for each of these three situations are in Fig. 6.8, presented from top to bottom as $\mathbf{P}(0)$ along x, y, and z, respectively. The situation in the bottom case is analogous to a LF μSR experiment, and the simulation correspondingly shows a set of signals consistent with this geometry. The other two cases correspond to transverse field (TF) μSR geometries, as indicated by the large precession amplitudes at small times.

It is interesting to note the long-time (asymptotic) values of each of the three signals from the bottom plot. The z signal starts at 1.0 and oscillates with a decaying amplitude. The asymptotic value of this signal is about 0.92, which is a reflection of the long-time value of the term b_z^2/b^2 in equation 6.11, averaged over the FLL unit cell area. The x signal of this plot has a long time value which is non-zero, and in fact has a value which is near 0.066. This value is the long-time value of the *correlation function* $< b_x b_z/b^2 >$ from equation 6.9, where the $< \dots >$ now indicate the average over the FLL area. The y signal also oscillates and decays with time, but its long-time value is zero. This means that in equation 6.10 the correlation function $< b_y b_z/b^2 >= 0$ as $t \to \infty$.

The average of the functions over the area of the FLL can be expressed mathematically as, *e.g.*:

$$\frac{b_x b_z}{b^2} \to < \frac{b_x b_z}{b^2} >= \int_A n(b) \frac{b_x b_z}{b^2} \, dA \qquad (6.12)$$

where b_x, b_z, and b are functions of the position within the FLL, $dA = dx \, dy$, and the arrow in the equation indicates that the term from equation 6.9 is taken over to an average over the FLL area. The lines on the graphs in Fig. 6.8 therefore really represent these averaged functions as functions of time, and should be thought of as such.

Another interesting interpretation of this long-time behavior is to visualize a net, ensemble average polarization vector and its projection onto the three directions. This vector begins, *e.g.*, aligned along z, and precesses on a cone which causes small oscillations in the x and y directions. As time progresses this cone becomes tilted off of the z axis and up toward the x axis by a small angle. The net polarization can then also be thought

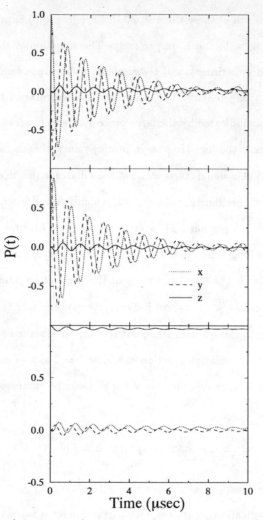

Figure 6.8: Shown from top to bottom are three simulations of μSR data from three different directions of the initial muon polarization. The top has $\mathbf{P}(0)$ along x, the middle along y, and the bottom along z. The other values of the parameters are: $B = 100\ G$, $\lambda = 256.5\ nm$, $\theta = 45°$, and $\Gamma = 25$.

to have shifted its orientation from along z to along this newly defined direction. This effect has been seen independently (and published earlier) by Riseman [17].

6.4.2 Finding Moments

The asymptotic behavior of the simulations noted above gives some information about particular moments of the field distributions at long times. However, this has arisen in an almost accidental fashion. It would therefore be useful to develop a method to actively extract further information (*i.e.* moments) from the simulations.

Our method of arriving at these moments involves manipulating the simulation data of the various plots in Fig. 6.8 using elementary mathematics, algebra, and Fourier transform techniques. For example, if we wish to find the moment $< b_x/b >$, we can do it by subtracting the z simulation curve of the center plot of Fig. 6.8 from the y simulation curve of the bottom plot. We then divide each element by 2, and take the Fourier sine transform over time. This results in a function of frequency, $< b_x(\omega)/b >$, which when integrated over frequencies yields a number for $< b_x/b >$.

How do we know this will give $< b_x/b >$? The answer requires inspection of equations 6.3 through 6.11. The equation corresponding to the y curve in the bottom plot is 6.10, while that corresponding to the z curve in the center plot is 6.6. Both of these equations contain the term b_x/b, and the curves for both are averaged over the same FLL area. It is therefore possible to do the subtraction of equation 6.6 from equation 6.10:

$$< \frac{b_y b_z}{b^2}(1 - \cos \omega t) > \quad + \quad < \frac{b_x}{b} \sin \omega t > -$$

$$\left[< \frac{b_z b_y}{b^2}(1 - \cos \omega t) > - < \frac{b_x}{b} \sin \omega t > \right]$$

$$= 2 < \frac{b_x}{b} \sin \omega t > \qquad (6.13)$$

Dividing by 2 and taking the sine Fourier transform gives:

$$\int_0^T < \frac{b_x}{b} \sin \omega t > \sin \omega' t \, dt \rightarrow \int_0^T \int_A n(b) \frac{b_x}{b} \sin \omega t \sin \omega' t \, dt \, dA \qquad (6.14)$$

Now switch the order of integrations:

$$\int_A n(b)\, dA \int_0^T \sin \omega t \sin \omega' t\, dt \tag{6.15}$$

The integral over time is approximately a delta function in frequency:

$$\int_0^T \sin \omega t \sin \omega' t = \frac{\sin(\omega - \omega')t}{2(\omega - \omega')} - \frac{\sin(\omega + \omega')t}{2(\omega + \omega')} \simeq \delta(\omega - \omega') \tag{6.16}$$

Substituting this in gives:

$$\int_A n(b) \frac{b_x}{b} \delta(\omega - \omega')\, dA = <\frac{b_x}{b}> n(b) \tag{6.17}$$

which is the moment $b_x/b\, n(b)$ of the field distribution. For the data shown in Fig. 6.8, the result using this procedure is in Fig. 6.9.

This plot represents the distribution $n(b)$ of the magnitude of the local fields *weighted* by the term b_x/b for each b within the distribution $n(b)$. It is *not* the distribution of b_x/b itself, which would most likely look much different. This idea of a weighted $n(b)$ distribution is central to understanding the concept of the moments, and is necessary for an understanding of what follows.

6.5 Interesting Moments

6.5.1 n(b)

Back in chapter 4 a theoretical procedure was discussed in which the distribution of the magnitude of the local field **b** could be obtained. This is of interest because it sheds light on the microscopic properties of the equilibrium FLL. It would therefore be useful if, in addition to this purely theoretical calculation, there were an experimental means to see this distribution. Using the techniques described above, such an experimental procedure is now described.

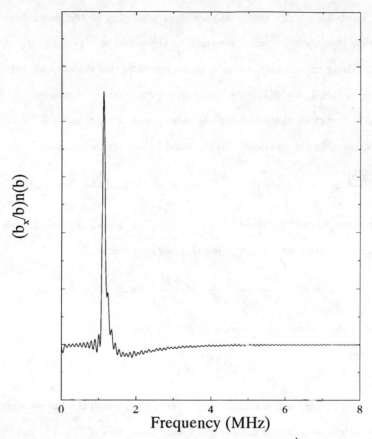

Figure 6.9: This is the plot resulting from the analysis deriving $\frac{b_x}{b} n(b)$ using the simulation data with parameters $B = 100\ G$, $\lambda = 256.5\ nm$, $\theta = 45^\circ$, and $\Gamma = 25$.

One first needs to acquire high statistics data according to the geometries of Fig. 6.8. This means that the data will correspond to the averaged versions of equations 6.3 through 6.11. Using these equations as a guide, we recognize that the averaged forms have terms which look like $< n(b) \frac{b_x^2 + b_y^2}{b^2} \cos \omega t >$, which appear in equations 6.5, 6.8, and 6.11. We first subtract off the long-time non-zero values, since these will be due to the non-sinusoidal terms in the equations. Next, we add these terms together:

$$< n(b) \frac{b_x^2 b_y^2}{b^2} \cos \omega t > + < n(b) \frac{b_y^2 + b_z^2}{b^2} \cos \omega t > + < n(b) \frac{b_z^2 + b_x^2}{b^2} \cos \omega t > \qquad (6.18)$$

All terms are averaged over the same FLL, so the $< \dots >$ are all the same. Also, $n(b)$ is exactly the same in each term. Recognizing this, we may write:

$$< n(b) \left(\frac{b_x^2 + b_y^2 + b_y^2 + b_z^2 + b_x^2}{b^2} \right) \cos \omega t > \qquad (6.19)$$

The term within the parentheses becomes $2b^2/b^2$, leaving:

$$< n(b) \, 2 \cos \omega t > \qquad (6.20)$$

Dividing by 2 and Fourier cosine transforming leaves $n(b)$, which is what we were after.

This analysis has been carried out using the simulation data of Fig. 6.8, and the result is shown in Fig. 6.10. Also shown is the result of the purely theoretical calculation of chapter 4 for comparison sake. The simulation result is not as clean and smooth as the integrated curve, but that is to be expected due to the lesser overall resolution afforded by the Fourier transform, as well as the "ringing" oscillations which occur as a result of the finite time range of the integration (recall the approximation to a gaussian above in equation 6.16).

The important thing to keep in mind here is that this a method which allows experimental extraction of the microscopic distribution of the magnitude of the local field b. What is most commonly done in μSR experiments is a Fourier transform of TF-μSR data at high applied fields. To look at what this actually means, we take, $e.g.$, equations 6.3

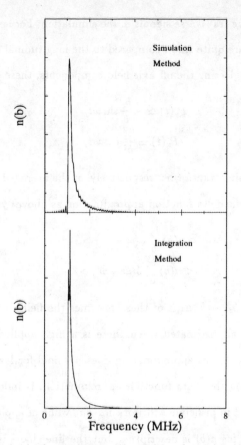

Figure 6.10: The top plot is the moment $n(b)$ resulting from this analysis of the simulation data. The bottom plot is the result for the same moment but with the integration method of last chapter – shown as a check of the method.

and 6.5, which detect the transverse signals in the simulation. For large applied fields, the components b_x and b_y are quite small compared to the longitudinal component b_z, which is nearly equal to b. Neglecting the off axis field components, these equations become:

$$P_y(t) \simeq -\frac{b_z}{b} \sin \omega t \qquad (6.21)$$

$$P_x(t) \simeq \frac{b_z^2}{b^2} \cos \omega t \qquad (6.22)$$

Taking the sine and cosine transforms, respectively, of the averaged forms of these equations (and using the same delta function approximation as above) yields:

$$< n(b)\, \frac{b_z}{b}\, \delta(\omega - \omega') > \qquad (6.23)$$

$$< n(b)\, \frac{b_z^2}{b^2}\, \delta(\omega - \omega') > \qquad (6.24)$$

Now, in the *limit* that $b_z \to b$, each of these becomes the field distribution $n(b)$. This limiting case is closely approximated when there is a high applied field, but, as mentioned earlier in chapter 4, this approximation does *not* hold for low applied fields. Our method, while also using the delta function approximation, is independent of the high field approximation, and in principle will work for *any* value of applied field.

The example above for $n(b)$ is descriptive, but the line shape is rather smooth. To show that the method of moment finding is in fact a good method, capable of reproducing more complicated forms, simulation data have been prepared corresponding to a field $B = 250\ G$, an effective penetration depth $\lambda = 120\ nm$, anisotropy parameter $\Gamma = 25$, and angle $\theta = 45^\circ$. Although these values do not correspond to any real material, the line shapes have more structure and therefore are a better test of the method. The results of the simulation and the integration are shown in Fig. 6.11, top and bottom, respectively.

This distribution has more structure, as indicated in the bottom plot of Fig. 6.11. There is a sharp rise at the minimum frequency, followed by a peak, and then a dip, then another peak, and then a long high frequency tail which is asymptotic to zero. The simulation does a good job of reproducing this structure, getting not only both peaks

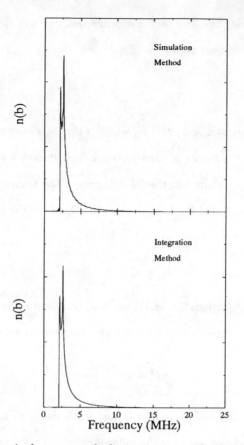

Figure 6.11: This figure is the same as the last, except now the parameters are: $B = 250\ G$, $\lambda = 120\ nm$, $\theta = 45°$, and $\Gamma = 25$. The integration result on the bottom shows more structure, which is well reproduced in the simulation distribution above.

correctly but also showing the long high frequency tail. We will show more of this type of simulation later for different moments.

6.5.2 $b_\perp^2 n(b)$

Another moment of interest is $< b_\perp^2 n(b) >$, which is the distribution of the magnitude of the field b weighted by the square of the transverse field at each b value within the FLL. This distribution can be obtained in the following way. The transverse field components are of course designated b_x and b_y, and therefore the square of the transverse component is simply:

$$b_\perp^2 = b_x^2 + b_y^2 \tag{6.25}$$

Upon inspection of equations 6.3 - 6.11, one sees this form contained in equation 6.11 as the right hand term. Subtracting off the time independent term from this equation leaves:

$$< \frac{b_x^2 + b_y^2}{b^2} \cos \omega t > \tag{6.26}$$

If we now take the Fourier cosine transform, and while doing so multiply each term by the field (frequency) squared, we have the following:

$$\int_0^T b^2 < \frac{b_x^2 + b_y^2}{b^2} \cos \omega t > \cos \omega' t \, dt = \tag{6.27}$$

$$\int_A n(b) b^2 < \frac{b_x^2 + b_y^2}{b^2} > \int_0^T \cos \omega t \cos \omega' t dt = < b_x^2 + b_y^2 > n(b) \tag{6.28}$$

The result of doing this calculation on the data of Fig. 6.8 is shown in Fig. 6.12 at the top. As a check on its validity, the integration method was again employed, but now modified to calculate the quantity $(b_x^2 + b_y^2) n(b)$ at each sub-cell of the FLL area, and weight appropriately as described above. The result from this method is shown in Fig. 6.12 in the bottom panel.

Once again, the integration method yields a much smoother result with finer detail. There is the rise at the minimum, a sharp peak, and then a high frequency tail which is

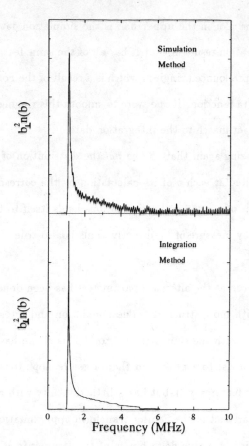

Figure 6.12: The top figure is the result for the moment $b_\perp^2\, n(b)$ using the simulation data from this chapter. The bottom plot is the same moment, but arrived at via the integration method developed in last chapter, and is shown to illuminate the validity of the method.

rather pronounced. The plot in the upper half is the simulation data, which reasonably well matches the integration result in that it has all of the same features of the shape. It does have some rather pronounced ringing, which is a result of the Fourier transformation approximation to a delta function. If one were to smooth this ringing out then the shape would be an almost exact match to the integration data.

It is worth emphasizing again that this is *not* the distribution of $b_\perp^2$ itself, but is the distribution $n(b)$ weighted at each b of its calculation by the corresponding value of $b_\perp^2$. The theoretical analysis of above for the simulations lends itself to this type of analysis and does not permit any measure of $b_\perp^2$ directly. This is also true of the other moments discussed here.

As before, another run at the alternate parameters has been done to show the reproducibility of shapes with more structure. The simulation and integration distributions are shown in Fig. 6.13, with the simulation on top. This shape has a sharp rise at the minimum, followed by a rise to a peak, and then a gentle slope toward zero. The simulation reproduces these features well, but has a little difficulty with ringing on the slope. This is most likely an artifact of the Fourier transform approximation mentioned above, in which the function is not a true delta but is of the form $\sin(a - b)/(a - b)$. Overall, however, the simulation reproduces the line shape well.

6.5.3 $b_x\,n(b)$

To find the moment $b_x\,n(b)$, one subtracts equation 6.6 from equation 6.10, which leaves:

$$< 2\frac{b_x}{b}\sin\omega t > \tag{6.29}$$

This data are then divided by two and Fourier sine transformed with each b as a coefficient in the transform. This leaves the moment $b_x\,n(b)$, which is the distribution of the x component of the microscopic fields within the FLL.

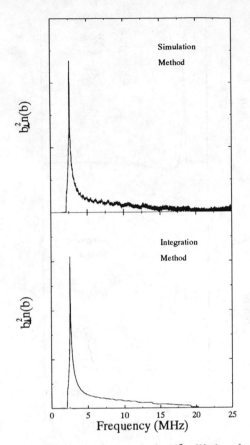

Figure 6.13: The top is the simulation result for $b_\perp^2 n(b)$ for the following parameters: $B = 250 \ G$, $\lambda = 120 \ nm$, $\theta = 45°$, and $\Gamma = 25$. The bottom is the corresponding integration result.

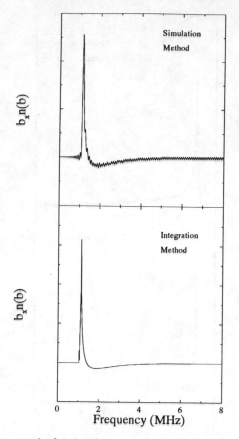

Figure 6.14: Shown on top is the simulation result for $b_x\, n(b)$ for the following parameters: $B = 100\ G$, $\lambda = 256.5\ nm$, $\theta = 45°$, and $\Gamma = 25$. The bottom plot is the corresponding integration method result via the methods of chapter 4.

The calculation for this moment is shown in Fig. 6.14, with the simulation result on top and the integration result on the bottom. For this case the parameters are: $B = 100\ G$, $\lambda = 256.5\ nm$, $\theta = 45°$, and $\Gamma = 25$. The simulation does a very good job in this case of reproducing the structure of the distribution. It is interesting to note for the theoretical moment that the area under the curve is zero, because the average over the FLL must be zero as stated in the theory. The simulation also does a good job in showing this feature.

The same distribution is again calculated for the alternate parameters of $B = 250\ G$, $\lambda = 120\ nm$, $\theta = 45°$, and $\Gamma = 25$. The plots are shown in Fig. 6.15, with the simulation

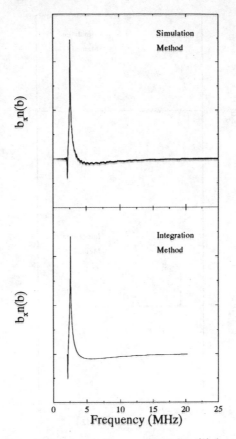

Figure 6.15: Shown on top is the simulation result for $b_x\, n(b)$ for the following parameters: $B = 250\ G$, $\lambda = 120\ nm$, $\theta = 45°$, and $\Gamma = 25$. The bottom plot is the corresponding integration method result via the methods of chapter 4.

again on top. There is a little more structure here, which is well represented in the simulation. The theoretical area is also zero, as it was above.

6.5.4 $b_y\, n(b)$

Another moment of interest is $b_y\, n(b)$, which is the distribution $n(b)$ weighted by the appropriate factor b_y at each b. This moment is found by subtracting equation 6.9 from 6.4, dividing this by 2, and then sine Fourier transforming while multiplying by the appropriate b_y. The result for this is shown in Fig. 6.16, again accompanied by the integration result.

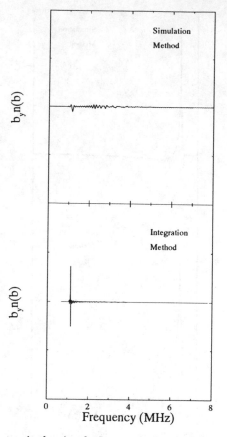

Figure 6.16: Shown on top is the simulation result for $b_y\, n(b)$ for the following parameters: $B = 100\ G$, $\lambda = 256.5\ nm$, $\theta = 45°$, and $\Gamma = 25$. The bottom plot is the corresponding integration method result via the methods of chapter 4.

These two curves are quite dissimilar. The integration result is largely zero, being broken up only by a very narrow up-down spike near 1 MHz. This makes sense if one recalls the surface of b_y, where there was almost perfect asymmetry within the FLL unit cell. At every point within the FLL cell there existed a b_y value whose opposite in sign could be found symmetrically across the FLL. It is therefore understandable to get a zero at all (most) points because the b_y factor which multiplies in the integral should sum to zero. This moment should integrate to zero via this line of reasoning, and it does.

The simulation result does not have the same shape. It starts out looking like the integration result, but after going negative it returns to zero very slowly as frequency increases. This obvious discrepancy is hard to explain, and must be due to more than the resolution afforded by the Fourier transform. It seems that the simulation is somehow not able to be selective enough in this case to insure proper cancellation of all of the b_y components. The moment does integrate to (near) zero, however, which will become important later. (The alternate simulation data look very similar to this, with the simulation not matching the integration very well. It, too, integrates to near zero, however.)

6.5.5 $b_z\, n(b)$

As a final example, the moment $b_z\, n(b)$ is calculated. This moment is found by subtracting equation 6.3 from equation 6.3, dividing by 2, and doing the sine transform while multiplying by each field within the transform – exactly as described above. This then results in the distribution $n(b)$ weighted by the microscopic z component of the fields b_z. The simulation and integration results for the $B = 100\ G$ case are shown in Fig. 6.17. These line shapes are similar to those for $n(b)$, where there is a large peak followed by a long, high frequency tail.

The comparison data are shown in Fig. 6.18 for the alternate field and parameter values. These look similar to the above $n(b)$ data for the alternate simulation data – Fig. 6.11. Here, as there, is more structure than in the main simulation case, with two peaks followed by a long tail. This time, as in most cases we have seen, the simulation does a very good job of producing the same result as the integration method.

6.5.6 Other Moments

It is possible to find other moments, such as $b_x\, b_z$, $b_z\, b_y$, $b_y^2 + b_x^2$, etc. These can be obtained via the methods discussed above, but will not be discussed because of their limited usefulness.

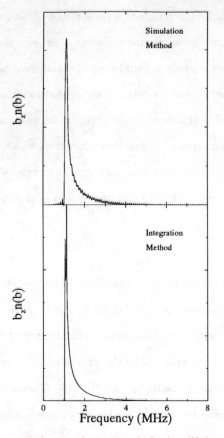

Figure 6.17: Shown on top is the simulation result for $b_z\,n(b)$ for the following parameters: $B = 100\ G$, $\lambda = 256.5\ nm$, $\theta = 45°$, and $\Gamma = 25$. The bottom plot is the corresponding integration method result via the methods of chapter 4.

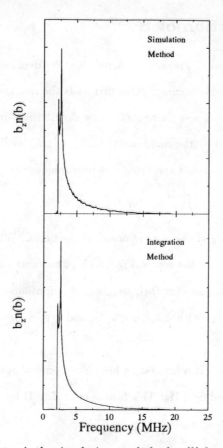

Figure 6.18: Shown on top is the simulation result for $b_z\, n(b)$ for the following parameters: $B = 250\ G$, $\lambda = 120\ nm$, $\theta = 45^\circ$, and $\Gamma = 25$. The bottom plot is the corresponding integration method result via the methods of chapter 4.

6.6   The Direction of B

Using the techiques of above it is possible to determine the direction of the average internal field $\mathbf{B} =< b >$ to reasonable accuracy. One first finds the moments $b_x\, n(b)$, $b_y\, n(b)$, and $b_z\, n(b)$ as described above. These moments are now distributions in frequency of the field distribution $n(b)$ weighted by the components b_x, b_y, and b_z, as discussed above. If one now takes each of these moments and integrates over frequencies (fields), one has (*e.g.*):

$$\int_{b_{min}}^{b_{max}} b_x\, n(b)\ db =< b_x > \tag{6.30}$$

which is the average value of the x component of the fields. What this amounts to is integrating out the curve on the top of Fig. 6.14. The same calculation can be done for both b_y and b_z, integrating over their weighted distributions in Figs. 6.16 and 6.17, respectively. One is then left with $< b_x >$, $< b_y >$, and $< b_z >$, which are the components of $\mathbf{B}$.

Why is this interesting? It is interesting because in general one knows only the direction of the external applied field $\mathbf{H}_a$. This field is related to $\mathbf{B}$ by the following equation:

$$\mathbf{H}_a = \mathbf{B} - 4\pi\mathbf{M} \tag{6.31}$$

where $\mathbf{M}$ is the magnetization of the material, and the equation is expressed in gaussian units [81]. It is therefore quite possible that if $\mathbf{M}$ is appreciable that the applied field $\mathbf{H}_a$ will not be parallel to the average internal field $\mathbf{B}$.

The theory developed in chapters 3 and 4 assumes one has a knowledge of the average field $\mathbf{B}$ in both magnitude and direction. The development in this chapter is built upon this theory and, therefore, this assumption. It is generally assumed in most theories and experiments that the applied and internal average fields are parallel. This method should now allow for an experimental check on the validity of this assumption.

Prior to discussing the results of this analysis, a brief discussion of some important points of the analysis will be presented. First, one must be sure to create a good set of simulation curves for the three, mutually perpendicular directions. The experimental analog is to make sure that all detectors are aligned and calibrated, that the muon initial polarization direction is reasonably well known, and that high statistics data are taken. Second, one must Fourier transform this data over the proper range of frequencies, keeping in mind that the lowest and highest frequencies possible to see are: $f_{min} = 1/T$, and $f_{max} = 1/2\Delta t$ (the Nyquist limit). Here T is the overall time that the time data are sampled, and Δt is the sampling time interval such that $T = N\Delta t$, where N is the total number of points in time space. Third, and last, is to make sure that a good integration algorithm is used to integrate over the moment distribution. The one used here is a Simpson's Rule algorithm adapted from ref. [97].

Before using the simulation data directly, the validity of the method was checked by using "dumb" data. This "dumb" data was generated from simple field distributions where the field was uniform throughout the FLL. The first case was for a field of 295.2 G directed along z. The second had this same field now oriented at $45°$ from each of the three axes, so that the component along each direction was 170.434 G. Finally, the signal from a single muon which stopped within the uniform field of the first case was used. The results of the moment analysis to these special cases are presented in Table 6.1.

It is apparent from this table that the method works rather well for these simple cases.

Next we apply the method to the real simulation data, like that of Fig. 6.8. We also apply it to the alternate data set, with $B = 250$ G, $\lambda = 256.5$ nm, $\theta = 45°$, and $\Gamma = 25$. Also, for one extra comparison, we apply the method to a situation where $B = 1000$ G, $\lambda = 256.5$ nm, $\theta = 45°$, and $\Gamma = 25$. The results are shown in Table 6.2.

These results are not as good as the simple cases, but that is to be expected due to the much more complicated magnetic field structure within the FLL. Most of the results

Initial Data (G)			Simulation Results (G)		
b_x	b_y	b_z	$< b_x >$	$< b_y >$	$< b_z >$
0	0	295.2	$.3 \times 10^{-11}$	$.3 \times 10^{-11}$	296.73
170.434	170.434	170.434	171.317	171.317	171.317
0	0	295.2	$.8 \times 10^{-11}$	$.8 \times 10^{-11}$	296.7

Table 6.1: These are the results of applying the moment method to simple field distributions directed along z, 45° to each axis, and again along z, for $B = 295.2$ G. The left side shows the components within the FLL, and the right shows the averaged values resulting from the analysis.

Simulation Results (G)			
$B \parallel z$	$< b_x >$	$< b_y >$	$< b_z >$
100	-0.0247	-0.0236	93.88
250	-0.026	0.074	243.77
1000	0.983×10^{-2}	-0.256×10^{-1}	999.73

Table 6.2: These are the results of applying the moment method to the different simulation data. The top results are for the $B = 100$ G, $\lambda = 256.5$ nm, $\theta = 45°$, and $\Gamma = 25$ case. The middle results are for the alternate simulation data: $B = 250$ G, $\lambda = 120$ nm, $\theta = 45°$, and $\Gamma = 25$. The last data are for $B = 1000$ G, $\lambda = 256.5$ nm, $\theta = 45°$, and $\Gamma = 25$.

are close to the proper values, and fall within about ±5%. Therefore if the actual **B** is off from the applied field $\mathbf{H}_a$ by more than this amount, we should be able to see it (assuming that we can get good μSR data with high statistics). Based on these results, it does not seem possible to detect any smaller deviation of **B** from $\mathbf{H}_a$ than this amount.

Chapter 7

Conclusions

This work set out to study the internal magnetic fields of the high temperature, anisotropic superconductors, particularly the compound $YBa_2Cu_3O_7$. Most of these new materials have T_c values (critical temperature – the onset of superconductivity) which are above the boiling point of liquid nitrogen. This fact has made their potential use very attractive due to the relative cheapness of liquid nitrogen compared to other coolants and refrigeration techniques. These newer superconductors, discovered in the late 1980's, have been some of the most intensly studied materials of all time. However, there is still much to be learned about them, especially at lower applied fields where many theoretical and experimental assumptions break down. This work has therefore been an attempt to better understand these materials from a theoretical point of view, and to provide a framework and the tools for an experiment to complement these studies.

To this end, the work first described the μSR techniques and how muons behave in the presence of a magnetic field. The data obtained with these techniques, and how it can be used to show the microscopic internal magnetic fields seen by the muons, was discussed in some detail.

Next, the isotropic and anisotropic London theories, which allow theoretical calculation of the microscopic magnetic fields within the mixed state of these superconductors, were discussed at length. Then, using the anisotropic London theory, a computer program was developed which calculates the microscopic magnetic fields at any point within the equilibrium (triangular) flux line lattice. The program accepted user-specified parameters for the values of the average field B magnitude, the effective penetration depth λ, the angle θ defining the angle between the crystal **c** axis and the average field **B** direction, and the anisotropy parameter Γ. The field components, including those which are transverse (or off-axis) to the average field **B** direction, were then calculated on a grid which subdivded the flux line lattice (FLL) unit cell.

Using these field values, the theoretical distributions of the magnetic fields were calculated. Distributions were calculated for all components of the fields, as well as for the magnitude of the field **b**. A study was then undertaken to determine the dependence of the fields and the distributions on the various parameters of the program, such as the average field B, the effective penetration depth λ, the anisotropy parameter Γ, and the angle θ between the crystal **c** axis and the average field **B**. The $1/\lambda^2$ dependence of the width of the field distributions was shown to hold, as well as the interesting dependence of the shape of the distribution on the magnitude of the average field. These results, and others, are discussed in chapter 5.

The last part of this work combined the knowledge gained from the study of muons as well as the results of the theoretical field data. That is, knowing the microscopic magnetic field within the FLL allowed a simulation of the muon behavior within the unit cell. A program was written which simulates the behavior of muons stopping uniformly within the FLL, calculating their polarization projection onto three mutually perpendicular "detectors." This simulation was done in an attempt to probe the microscopic magnetic field components which are transverse (or, off-axis) to the direction of the average magnetic field **B**. These transverse fields arise naturally from the theory when the anisotropy is taken into account, and are not due to pinning or other mechanism external to the fields.

The simulation results show a depolarization of the muon spins for all three directions – meaning that there is field inhomogeniety within the FLL and proving (at least at some level) that these "off axis" fields exist. (Incidentally, to the best of this author's knowledge, this simple experiment has yet to be done.)

The final endeavor of this work was to use the results of the simulations to extract further information about the magnetic fields within the FLL. A technique was then developed to take the simulation data, manipulate it using both simple math and Fourier analysis, and extract information which we call *moments* of the field distributions. One useful moment, $n(b)$, is the distribution of the magnitude of the microscopic magnetic field **b** within the FLL. The only approximation in this method is that of the upper limit of a Fourier integral being finite and not infinite, which introduces an approximate delta function as opposed to a real one. This approximation, however, is common to all numerical Fourier transform techniques, and should still yield good, quantitative results. Another useful result of the analysis is that the direction of the average internal magnetic field **B** may be determined to reasonable accuracy. This is useful in that its direction may in general be different than that of the applied field $\mathbf{H}_a$.

The types of superconductors to which this theory naturally applies – single crystals of high quality $YBa_2Cu_3O_7$ – are now being produced with little twinning and defects. It is in these materials where one hopes to best apply the results of the London theory, since they most closely approximate the theoretical assumptions. It is hoped that some day the experimental analog of this theoretical study may be pursued in hopes of shedding more light onto the still confusing data and theories which surround these materials. This may be stated with confidence because, at the time of this writing, a grant based largely on this work has been funded for these experiments.

Bibliography

[1] J.G. Bednorz and K.A. Müller. *Z. Phys. B*, **64**:189, 1986.

[2] R.L. Garwin, L.M. Lederman, and M. Weinrich. Observations of the failure of conservation of parity and charge conjugation in meson decays: The magnetic moment of the free muon. *Phys. Rev.*, **105**:1415, 1957.

[3] J.I. Friedman and V.L. Telegdi. *Phys. Rev.*, **106**:1290, 1957.

[4] T.D. Lee and C.N. Yang. *Phys. Rev.*, **98**:1501, 1955.

[5] T.D. Lee and C.N. Yang. *Phys. Rev.*, **104**:254, 1956.

[6] J. Chappert. In J. Chappert and R.I. Grynszpan, editors, *Muons and Pions in Materials Research*, page 35. North-Holland, 1984.

[7] A. Schenck. *Muon Spin Rotation Spectroscopy*. Adam Hilger Ltd., Boston, 1985.

[8] J.H. Brewer and K.M. Crowe. Advances in muon spin rotation. In J.D. Jackson *et al.*, editor, *Annual Review of Nuclear and Particle Science*, volume **28**, page 239. Palo Alto Reviews, Inc., 1978.

[9] W.J. Kossler, C.M. Benesch, and A.J. Greer. A surface muon beam at CEBAF. Internal Report, 1993.

[10] M. Goldhaber, L. Grodzins, and A. Sunyar. *Phys. Rev*, **109**:1015, 1958.

[11] Particle Data Group. *Particle Properties Data Booklet*. North-Holland, 1990.

[12] F. Scheck. page 1. Bürgenstock Meeting, 1971.

[13] L.B. Okun. *Weak Interactions of Elementary Particles*. Pergamon, Oxford, 1965.

[14] S.F.J. Cox. In J. Chappert and R.I. Grynszpan, editors, *Muons and Pions in Materials Research*, page 185. North-Holland, 1984.

[15] G.W. Ford and C.J. Mullin. *Phys. Rev.*, **108**:1415, 1957.

[16] J. Chappert and R.I. Grynszpan, editors. *Muons and Pions in Materials Research*. North-Holland, 1984.

[17] Tanya M. Riseman. *μSR Measurement of the Magnetic Penetration Depth and Coherence Length in the High-Tc Superconductor $YBa_2Cu_3O_{6.95}$*. PhD thesis, University of British Columbia, 1993.

[18] R. Kubo and T. Toyabe. A stochastic model for low field resonance and relaxation. In R. Blinc, editor, *Magnetic Resonance and Relaxation*, page 810. North Holland, Amsterdam, 1967.

[19] K.G. Petzinger and S.H. Wei. *Hyperfine Interactions*, 1983.

[20] M. Celio and P.F. Meier. *Phys. Rev. B*, **27**:1908, 1983.

[21] H. Kammerlingh Onnes. *Leiden Comm.*, pages 120b,122b,124c, 1911.

[22] Neil W. Ashcroft and N. David Mermin. *Solid State Physics*, chapter 34. W.B. Saunders Co., 1976.

[23] W. Meissner and R. Ochsenfeld. *Naturwissenschaften*, **21**:787, 1933.

[24] D.R. Harshman, G. Aeppli, E.J. Ansaldo, B. Batlogg, J.H. Brewer, J.F. Carolan, R.J. Cava, M. Celio, A. Chaklader, W.N. Hardy, S.R. Kreitzman, G.M. Luke, D.R. Noakes, and M. Senba. *Phys. Rev. B*, **36**:2386, 1987.

[25] Dong-Ho Wu and S. Sridar. *Phys. Rev. Lett.*, **65**:2074, 1990.

[26] V.J. Emery. *Phys. Rev. Lett.*, **58**:2794, 1987.

[27] W.N. Hardy, D.A. Bonn, D.C. Morgan, Ruixing Liang, and Kuan Zhang. *Phys. Rev. Lett.*, **70**:3999, 1993.

[28] S. Sridar, D.H. Wu, and W. Kennedy. *Phys. Rev. Lett.*, **63**:1873, 1989.

[29] J.E. Sonier *et al. Phys. Rev. Lett.*, **72**:744, 1994.

[30] D.A. Wollman, D.J. Van Harlingen, W.C. Lee, D.M. Ginsberg, and A.J. Leggett. *Phys. Rev. Lett.*, **71**:2134, 1993.

[31] A. Balatsky and E. Abrahams. *Phys. Rev. B*, **45**:13125, 1992.

[32] J. Bardeen, L.N. Cooper, and J.R. Schrieffer. Theory of superconductivity. *Phys. Rev.*, **108**:1175, 1957.

[33] Michael Tinkham. *Introduction to Superconductivity*. McGraw-Hill, 1975.

[34] P.G. deGennes. *Superconductivity of Metals and Alloys*. Addison-Wesley, Redwood City, CA, 1966.

[35] M.K. Wu *et al. Phys. Rev. Lett.*, **58**:908, 1987.

[36] A.B. Pippard. *Proc. Royal Soc.*, **A216**:547, 1953.

[37] L.L. Campbell, M.M. Doria, and V.G. Kogan. *Phys. Rev. B*, **38**:2439, 1988.

[38] C. W. Chu *et al. Phys. Rev. Lett.*, **58**:405, 1987.

[39] D.R. Harshman and A.P. Mills. *Phys. Rev. B*, **45**:10684, 1992.

[40] V.L. Ginzburg and L.D. Landau. *Zh. Eksperim. i Teor. Fiz.*, **20**:1064, 1950.

[41] F. London and H. London. *Proc. Roy. Soc. (London)*, **A149**:71, 1935.

[42] V.G. Kogan. *Phys. Lett.*, **85A**:298, 1981.

[43] Max von Laue. *Annalen der Physik*, **6**:31, 1948.

[44] D.E. Farrell *et al. Phys. Rev. Lett.*, **61**:2805, 1988.

[45] D.E. Farrell *et al. Phys. Rev. Lett.*, **63**:782, 1989.

[46] S.L. Thiemann, Z. Radović, and V.G. Kogan. *Phys. Rev. B*, **39**:11406, 1989.

[47] A.V. Balatskiĭ, L.I.Burlachov, and L.P. Gor'kov. *Sov. Phys. JETP*, **63**:866, 1986.

[48] T. Krzyston and P. Wróbel. *Phys. Stat. Sol.*, **145**:K41, 1988.

[49] V.G. Kogan. *Phys. Rev. B*, **38**:7049, 1988.

[50] E.H. Brandt. *Journal of Low Temperature Physics*, **73**:355, 1988.

[51] E.H. Brandt and A. Seeger. *Advances in Physics*, **35**:189, 1986.

[52] D.R. Harshman *et al.* private communication.

[53] D.R. Harshman, E.H. Brandt, A.T. Fiory, M. Inui, D.B. Mitzi, L.F. Schneemeyer, and J.V. Waszczak. *Phys. Rev. B*, **47**:2905, 1993.

[54] I.V. Grigorieva, L.A. Gurevich, and L.Ya. Vinnikov. *Physica C*, **195**:327, 1992.

[55] M. Yethiraj *et al. Phys. Rev. Lett.*, **70**:857, 1993.

[56] B. Keimer, F. Dogan, I.A. Aksay, R.W. Erwin, J.W. Lynn, and M. Sarikaya. *Science*, **262**:83, 1993.

[57] D.M. Paul, E.M. Forgan, R. Cubitt, S.L. Lee, M. Yethiraj, and H.A. Hook. *Physica B*, **192**:70, 1992.

[58] P.L. Gammel, D.J. Bishop, J.P. Rice, and D.M. Ginsberg. *Phys. Rev. Lett.*, **68**:3343, 1992.

[59] W.K. Kwok, S. Fleshler, U. Welp, V.M. Vinokur, J. Downey, G.W. Crabtree, and M.M. Miller. *Phys. Rev. Lett.*, **69**:3370, 1992.

[60] B.I. Ivlev and N.B. Kopnin. *Phys. Rev. B*, **44**:2747, 1991.

[61] B.I. Ivlev *et al. Phys. Rev. B*, **43**:2896, 1991.

[62] V.G. Kogan *et al. Phys. Rev. B*, **42**:2631, 1990.

[63] A.I. Buzdin and A. Yu Siminov. *JETP Lett.*, **51**:191, 1990.

[64] A.M. Grishin *et al. Sov. Phys. JETP*, **70**:1930, 1990.

[65] L.L. Daemen, L.J. Campbell, and V.G. Kogan. *Phys. Rev. B*, **46**:3631, 1992-II.

[66] C.A. Bolle, P.L. Gammel, D.G. Grier, C.A. Murray, D.J. Bishop, D.B. Mitzi, and A. Kapitulnik. *Phys. Rev. Lett.*, **66**:112, 1991.

[67] A. Sudbø and E.H. Brandt. *Phys. Rev. Lett.*, **68**:1758, 1992.

[68] L.L. Daemen, L.J. Campbell, A.Yu. Siminov, and V.G. Kogan. *Phys. Rev. Lett.*, **70**:2948, 1993.

[69] W.E. Lawrence and S. Doniach. In E. Kanda, editor, *Proceedings of the 12th International Conference on Low Temperature Physics*, page 361, 1970.

[70] R. Cubitt *et al. Physica C*, **213**:126, 1993.

[71] S.L. Lee *et al. Phys. Rev. Lett.*, **71**:3862, 1993.

[72] S. Senoussi, F. Mosbah, O. Sarrhini, and S. Hammond. *Physica C*, **211**:288, 1993.

[73] B. Batlogg. *Physics Today*, **44**:44, 1991.

[74] Yethiraj *et al. Phys. Rev. Lett.*, **71**:3019, 1993.

[75] R. Liang, P. Dosanjh, D.A. Bonn, D.J. Baar, J.F. Carolan, and W.N. Hardy. *Physica C*, **195**:51, 1992.

[76] D.R. Harshman, A.T. Fiory, and R.J. Cava. *Phys. Rev. Lett.*, **66**:3313, 1991.

[77] P. Carretta and M. Corti. *Phys. Rev. Lett.*, **68**:1236, 1992.

[78] B. Pümpin *et al. Phys. Rev. B*, **42**:8019, 1990.

[79] K.G. Petzinger and G.A. Warren. *Phys. Rev. B*, **42**:2023, 1990.

[80] H. Krakauer. private communication.

[81] J.D. Jackson. *Classical Electrodynamics*. John Wiley and Sons, second edition, 1975.

[82] M. Celio, T.M. Riseman, R.F. Kiefl, J.H. Brewer, and W.J. Kossler. *Physica C*, **153-155**:753, 1988.

[83] T.M. Riseman *et al. Physica C*, **162-164**:1555, 1989.

[84] A.D. Sidorenko, V.P. Smilga, and V.I. Fesenko. *Physica C*, **166**:167, 1990.

[85] A.D. Sidorenko, V.P. Smilga, and V.I. Fesenko. *Hyperfine Interactions*, **63**:49, 1990.

[86] M. Hein, G. Müller, H. Piel, L. Ponto, M. Becks, U. Klien, and M. Peiniger. *J. Appl. Phys.*, **66**:5940, 1989.

[87] M. Hein. *Hochfrequenz-Eigenschaften Granularer Hochtemperatur-Supraleiter*. PhD thesis, BUGH Wuppertal, May 1992.

[88] Y.J. Uemura and *et al.* In D. Givord, editor, *Proceedings of the International Conference on Magnetism*, volume 49, pages C8–2087. Journal de Physique, 1988.

[89] B.J. Chen, M.A. Rodriguez, S.T. Misture, and R.L. Snyder. *Physica C*, **198**:118, 1992.

[90] W.J. Kossler *et al. Phys. Rev. B*, **35**:7133, 1987.

[91] C.S. Jee and *et al. J. Supercond.*, **1**:63, 1988.

[92] R.J. Cava and *et al. Phys. Rev. Lett.*, **58**:1676, 1987.

[93] M. Weber, P. Birrer, F.N. Gygax, B. Hitti, E. Lippelt, M. Maletta, and A. Schenck. *Hyperfine Interactions*, **63**:93, 1990.

[94] P. Birrer, D. Cattani, J. Cors, M. Decroux, Ø. Fischer, F.N. Gygax, B. Hitti, E. Lippelt, A. Schenck, and M. Weber. *Hyperfine Interactions*, **63**:103, 1990.

[95] W.K. Dawson, K. Tibbs, S.P. Weathersby, C. Boekema, and K. Chan. *J. Appl. Phys.*, **64**:5809, 1988.

[96] W.K. Dawson and *et al. Hyp. Int.*, **63**:219, 1991.

[97] W.H. Press, B.P. Flannery, S.A. Teukolsky, and W.T. Vetterling. *Numerical Recipes: The Art of Scientific Computing.* Cambridge University Press, 1986.

List of Tables

List of Figures

159

Springer-Verlag
and the Environment

We at Springer-Verlag firmly believe that an international science publisher has a special obligation to the environment, and our corporate policies consistently reflect this conviction.

We also expect our business partners – paper mills, printers, packaging manufacturers, etc. – to commit themselves to using environmentally friendly materials and production processes.

The paper in this book is made from low- or no-chlorine pulp and is acid free, in conformance with international standards for paper permanency.

Lecture Notes in Physics

For information about Vols. 1–409
please contact your bookseller or Springer-Verlag

New Series m: Monographs

Vol. m 1: H. Hora, Plasmas at High Temperature and Density. VIII, 442 pages. 1991.

Vol. m 2: P. Busch, P. J. Lahti, P. Mittelstaedt, The Quantum Theory of Measurement. XIII, 165 pages. 1991.

Vol. m 3: A. Heck, J. M. Perdang (Eds.), Applying Fractals in Astronomy. IX, 210 pages. 1991.

Vol. m 4: R. K. Zeytounian, Mécanique des fluides fondamentale. XV, 615 pages, 1991.

Vol. m 5: R. K. Zeytounian, Meteorological Fluid Dynamics. XI, 346 pages. 1991.

Vol. m 6: N. M. J. Woodhouse, Special Relativity. VIII, 86 pages. 1992.

Vol. m 7: G. Morandi, The Role of Topology in Classical and Quantum Physics. XIII, 239 pages. 1992.

Vol. m 8: D. Funaro, Polynomial Approximation of Differential Equations. X, 305 pages. 1992.

Vol. m 9: M. Namiki, Stochastic Quantization. X, 217 pages. 1992.

Vol. m 10: J. Hoppe, Lectures on Integrable Systems. VII, 111 pages. 1992.

Vol. m 11: A. D. Yaghjian, Relativistic Dynamics of a Charged Sphere. XII, 115 pages. 1992.

Vol. m 12: G. Esposito, Quantum Gravity, Quantum Cosmology and Lorentzian Geometries. Second Corrected and Enlarged Edition. XVIII, 349 pages. 1994.

Vol. m 13: M. Klein, A. Knauf, Classical Planar Scattering by Coulombic Potentials. V, 142 pages. 1992.

Vol. m 14: A. Lerda, Anyons. XI, 138 pages. 1992.

Vol. m 15: N. Peters, B. Rogg (Eds.), Reduced Kinetic Mechanisms for Applications in Combustion Systems. X, 360 pages. 1993.

Vol. m 16: P. Christe, M. Henkel, Introduction to Conformal Invariance and Its Applications to Critical Phenomena. XV, 260 pages. 1993.

Vol. m 17: M. Schoen, Computer Simulation of Condensed Phases in Complex Geometries. X, 136 pages. 1993.

Vol. m 18: H. Carmichael, An Open Systems Approach to Quantum Optics. X, 179 pages. 1993.

Vol. m 19: S. D. Bogan, M. K. Hinders, Interface Effects in Elastic Wave Scattering. XII, 182 pages. 1994.

Vol. m 20: E. Abdalla, M. C. B. Abdalla, D. Dalmazi, A. Zadra, 2D-Gravity in Non-Critical Strings. IX, 319 pages. 1994.

Vol. m 21: G. P. Berman, E. N. Bulgakov, D. D. Holm, Crossover-Time in Quantum Boson and Spin Systems. XI, 268 pages. 1994.

Vol. m 22: M.-O. Hongler, Chaotic and Stochastic Behaviour in Automatic Production Lines. V, 85 pages. 1994.

Vol. m 23: V. S. Viswanath, G. Müller, The Recursion Method. X, 259 pages. 1994.

Vol. m 24: A. Ern, V. Giovangigli, Multicomponent Transport Algorithms. XIV, 427 pages. 1994.

Vol. m 25: A. V. Bogdanov, G. V. Dubrovskiy, M. P. Krutikov, D. V. Kulginov, V. M. Strelchenya, Interaction of Gases with Surfaces. XIV, 132 pages. 1995.

Vol. m 26: M. Dineykhan, G. V. Efimov, G. Ganbold, S. N. Nedelko, Oscillator Representation in Quantum Physics. IX, 279 pages. 1995.

Vol. m 27: J. T. Ottesen, Infinite Dimensional Groups and Algebras in Quantum Physics. IX, 218 pages. 1995.

Vol. m 28: O. Piguet, S. P. Sorella, Algebraic Renormalization. IX, 134 pages. 1995.

Vol: m 30: A. J. Greer, W. J. Kossler, Low Magnetic Fields in Anisotropic Superconductors. VII, 161 pages. 1995.